The Life Story of
Hubert Dielen
(1855-1926)

an autobiography

Cover–Photograph of Hubert Dielen in his late 60's and Dielen Family photograph taken in August 1937

The original book in Dutch was compiled by Jan Dielen.
The original title in Dutch is, "De lotgevallen van een Venlosche Slager"
Translated as: "The Adventures of a Venlo Butcher"

Translated into English by Tina Reininger
Transcribed and edited by Mike Reininger

First Edition in English Printed March 2005
Second Edition in English Printed September 2005
Third Edition in English Printed August 2006

Fourth Edition in English Printed April 2007

ISBN: 978-0-9772896-1-5
LCCN: 2007900974

Printed in the United States of America

Contents

Prologue

This book is the life story of my grandfather Hubert Theodor Dielen, who was born in Straelen, Germany on the 28th of January 1855 and died in Venlo on the 6th of April 1926. It took him 12 notebooks to write down the life experiences that he thought were worth remembering. There are other anecdotal stories through the family and others in the city of Venlo but we'll not discuss them in this book.

This story begins in the year 1820. Europe was still recovering from the ravages of war left by Napoleon. The energetic King William I had been ruling the United Kingdom of the Netherlands since 1814 and was doing everything he could to get his impoverished country back on its feet.

To give you an idea what condition the country was in, the three provinces: Overijssel, Gelderland and Limburg did not have one paved road. There were only cart tracks that connected cities to villages. What that meant for winter transportation one can only imagine. Elsewhere, transportation by steam engine train was yet to occur for another 19 years. But the most important thing was that England had a definite advantage over Holland in the world (industrial) ranking because the industrial revolution in England was in full swing before 1800, whereas in the Netherlands, they had to wait until 1850 before even a few steam machines made their appearance.

Germany at that time was still only a loose confederation of 34 kingdoms, dukedoms/duchies, diocese, etc. It was not until 46 years later, in 1866, that Otto Von Bismarck was to appear and unite the loose confederation into one country. In 1839, the province of Limburg formed its current (1995) geographic shape. In 1830, Limburg was part of the Belgian Independence movement (from Protestant Netherlands) but by 1839, the state powers that then existed concluded that (eastern) Limburg [in contrast to Belgian (western) Limburg] would be annexed by the Netherlands. (Eastern) Limburg was also a part of the German confederation until 1866 when it then seceded. Maastricht and Venlo, up till 1867, were gated defensive forts.

Jan Dielen

(Editor's note: I recommend the reader have an historical atlas at his/her side to further understand the time period and locations.)

A sample of Hubert Dielen's handwriting from the Foreword of his autobiography.

Voorwoord!

Door veele Zorgen en te harde Anstrenging.
der laatste Maanden werd ik in Juli 1924 door
een Zenuwen lijden aangegrepen en kan niet
meer eeten. Doktoren Lomknecht die mijn meer-
maale onderzocht verklaarde, dat Hart en
Longen in beste Conditie waren, maar de
overspannene Zenuwen op mijn Maag werkten
en verordende, uitsluitend Melk met Eieren te
gebruiken en daarlijk Rust en nach een Rust
moest neemen, en die Ziekte Maanden lang kan
duuren. Daar ik nu mijn geheel leeven haard
gewerkt werd mijn dat niets doen, wenschelijk
lijk. Met leesen had ik al een paar Maanden

daar gebracht en nu besloot ik eens mijne
Ervaaringen van mijn veel bewogen leeven
op Papier te zetten. Het is te begrijpen dat
iemand met een leeftijd van 70 Jaaren veel
heeft meegemaakt, en men niet alles op
Papier kan zetten, maar zal ik trachten
zoo veel mogelijk de Hoofdpunten naar waar-
heid weer te geeven om niet aan te vergeeten
hijt te onttrekken.

De Schrijver.

Foreword

Due to the anxieties and exhaustion of recent months, I became in July 1925 afflicted with a nervous condition and have lost my appetite. Doctor Janknecht, who examined me several times, declared that my heart and lungs were in excellent condition but since my overtaxed nerves affected my stomach, I was ordered on a diet of only milk and eggs. In addition to that, I had to rest and rest some more because this illness could take a long time. Since I had worked hard my whole life, I found doing nothing horrible. I spent a couple months reading but now I've decided to put the experiences from my eventful life on paper. It's understandable that a 70-year-old person has gone through a lot and couldn't put it all on paper, but I shall attempt as much as possible to accurately retrace the main events so they won't be left to oblivion.

The Writer
(Hubert Dielen)

Local Towns and Villages in the Vicinity of Venlo Around 1875

I. Grandparents & Parents

It was a raw autumn morning in the year 1820, when a cold northwestern wind swept up whitecap waves on the Maas River. The waves battered against the chain-anchored ferryboat at the state ferry dock, *de Staai,* which was across the river from Venlo. From the direction of the village Grubbenvorst approached a fully loaded covered wagon drawn by an old horse. Alongside walked a tall, thin, young man of about 18 years of age with his hands in his pant pockets.

The ferryman, who just got out of bed and was still wiping the sleep from his eyes, asked, "Where does this party have to go this early in the morning and where do you come from?" To which the wagon guide replied in a gruff tone, "That's none of your business. We just want to be ferried over." The ferryman took umbrage and said, "You may be right at that. But for your information, the state ferry doesn't begin until 8 a.m. Nor am I even sure I want to cross with all this river turbulence."

Mathijs Dielen was sitting on a bundle of straw in the covered wagon with his wife and two children and was the owner of the furnishings. Hearing the contention, he jumped from the wagon to interject, "Ferryman, I'll bet you didn't sleep very well," he said genially, "and crawled out of the nest a little too early, eh? But listen, we come from Broekhuizen and want to go to Venlo if possible." The ferryman then asked, "What do you want to do there?" "Well," said Mathijs, "you know that on the 1st of January, Venlo will begin levying a duty on incoming meat, and you also know that Venlo depends upon the surrounding village butcheries, of which I belong, for their meat supplies. I'm no longer disposed to enduring the cold and poverty of village life and that's why I'm going with my wife and children to try my luck in Venlo." The ferryman disagreed with Mathijs and replied, "In a village, no one has ever died of starvation, but in a large enclosed city, as Venlo, no one cares whether you live or die. But alright, give me 6 stivers and get on the ferry with your belongings."

Mathijs' wife was a little woman and kept the family fortune in a blue linen, money-satchel. She was clearly able to hear the conversation as she emerged from the covered wagon. But she now hesitated to give the 6 stivers and even considered returning. But the wagon guide interrupted, "For that amount of money, I won't take you back anymore and if you don't hurry up, I'll turn the wagon around and go home." The poor frail woman, shivering from the cold, then charily opened her small money-satchel and gave the 6 stivers to the ferryman. The wagon then proceeded onto the ferryboat. Within 30 minutes, they arrived at the Maaspoort (poort means gate or gateway; hence, gateway to the Maas River). The gatekeeper demanded the toll tax but after he inspected their possessions, he waived the tax and they entered free of charge.

They arrived at Gasthuisstraat (Hospital Street) #375, where Mathijs had a small house rented named *The Sieve.* The little woman with her two small children, Johan and Wijnand, got off the wagon and although cold and numb, they looked clean and presentable. The neighbors came out of their homes with their usual curiosity. The next-door neighbor took pity on the family and invited them into her house so the family could warm themselves beside the stove. The neighbor then presented them with a cup of coffee and a large sandwich. All the while, Mathijs and the wagoner were diligently unloading the possessions and soon the kind neighbors were giving a helping hand. Thus was Mathijs' wife able to set up her new household before it became dark.

Mathijs wasn't all that large of built or strong, but he was, however, diligent and hard working, as was his wife who also possessed great Christian virtues. Due to her cleanliness, people preferred shopping at *The Sieve* when buying their meat, which enabled the store to progressively prosper. It wasn't long before *The Sieve* was too small. So in 1824, they rented a larger house at the intersection of St. Nicholas and Vleeschstraat (Meat Street). Johan helped his father wherever he could.

But the family's luck wasn't to last, for in 1825, Mathijs became ill and died soon thereafter, on January 4th, 1826. A butcher, named Leviedekus, from Jodenstraat (Jewish Street) offered his help, which naturally, the widow gratefully accepted. Leviedekus did the buying of the cows and Johan did the slaughtering in addition to helping, with the assistance of Wijnand, Leviedekus clean his residence. The Dielen's were thus able to sell half a cow and live adequately from the profit.

On a Monday evening, Leviedekus went to the widow and said he bought two inexpensive animals (cows) in the land of Cuyk but he must take them that coming Wednesday and must have half of the amount (of the cost price) tomorrow evening without fail. The widow assembled, with great effort, the required sum of money. So, when Leviedekus showed up Tuesday evening, the little woman was able to show, with pride and joy, the sought after money and handed the sum to Leviedekus. The following day, Wednesday, Johan went in the early evening to see if Leviedekus had returned, but the doors were locked. He tried later that evening but again no answer, then came a long sleepless night. Early in the morning, Johan was right back at the Leviedekus house on Jodenstraat (Jewish Street), and knocked several times to no avail. A neighbor woman came out (of her house) and said to Johan, "That dirty jew has flown the coop last night with all his belongings to (New York via) Antwerp and I'll bet he took you people for everything you're worth." Johan was in shock and couldn't speak. He burst into tears and cried bitterly.

When Johan returned home, he could barely tell his mother what he had heard. When she heard the story, she collapsed on the sofa weeping and sobbing. The children became terrified and ran for help to their neighbor Linders, a goldsmith, who immediately came to the house. The doctor was summoned and then the old venerable Vicar Dean also showed up. The whole house became packed with neighbors. Johan had to explain what had happened because, although his mother had somehow regained her senses, she was still crying and too upset to give a coherent story. Needless to say that everyone was condemning the way of that jew, but the damage had been done and now help was what was needed to soften the blow. The good Vicar Dean comforted the widow and her children and departed from the home with Mr. Linders. At Linders' home was a deep discussion about what could be done and it was decided to offer the widow one of their small rooms upstairs. She was then able to sew and launder to earn some money. Johan got a job at a busy bakery owned by Brienkman in Vleeschstraat (Meat Street). The bakery was later taken over by Frans Bloem. Wijnand got a job with a well-known mason in Helden (3 miles northwest of Kessel, Limburg).

It was 1830 when the Belgians entered Venlo. Baker Brienkman was very busy day and night baking and supplying breads. Johan Dielen, who had just been hired at Brienkman's, had to work extremely hard. This was more than Johan could endure. He was not a strong boy. He soon started to spit up blood and had to return to his poor distraught mother. He recuperated soon thereafter. Then, an old acquaintance of the family by the name of Benz, a butcher from Straelen, Germany, offered to take Johan with him as a butcher. The mother agreed under the condition that Johan could come home every weekend with his small earnings and a good piece of meat. This was how Mother Dielen was able to sustain herself in addition to the small income she received from taking in laundry and ironing work. That worked for a while until eventually she felt so overwhelmed she died the following year in 1832. Johan stayed in Straelen where he worked very hard and was well accepted.

While working for 14 years in the Benz butcher shop, Johan met and married Regina Brockmann on October 23, 1845. She was the daughter of Peter Paul Brockmann, a weapon maker, but was raised by her father's brother, who was the local magistrate. Uncle Brockmann gave the young couple a house and two acres of land (in Straelen) on Venloschestraat (Venlo Street) #25. They were then able to start their own small butchery.

Johan was not stocky or well-built, but he was a very hard worker. Regina was a model housewife. She also took in some boarders and hired a robust housekeeper, while Johan had hired an orphan to help him. Meanwhile, Uncle Brockmann's wife, the former S. Clever from Blitterswijk, died soon thereafter and Uncle Brockmann moved in with Johan and Regina. Brockmann then made Regina his sole heir, which prompted a lot of jealousy and resentment from his relatives.

Johan and Regina were childless the first 2 years of their marriage but then were blessed with a daughter and four sons. Their names were Hendrika, Louis, Johan, Hubert, and Philip.

The family was exemplary. Every morning, Regina and her housekeeper would go to the first High Mass at 6 a.m., Johan, my father, would go to the 7 a.m. Mass, and the late Uncle Brockmann attended the most Holy Mass at 8 a.m. After supper every evening, the household would then gather to pray the rosary. This brought considerable peace to the family and the children were reared in much piety.

Johan offered his brother, Wijnand, to come and live with him. Wijnand accepted the offer and assisted with the butchering and the farm work. Later on, Wijnand married Regina's cousin. They had one daughter but the three of them didn't live very long thereafter. Marie (Mieke Tante or Aunt Mary) who was the sister of Johan and Wijnand, worked as a housekeeper for the clergy in the area. When she was a little over her 50th birthday, the pastor in Kessel (Limburg) passed away. Marie was then unemployed but she had saved enough, in addition to an inheritance, to retire comfortably. Soon thereafter, she got married to an old widower named Bloemers. It wasn't long before he had squandered all her savings. She died nearly in poverty a couple of years later in January 1900. I'll talk more about this later.

The children of Johan & Regina grew up nicely. The daughter, Hendrika, became a real support for her mother. Then, in 1877, she got married to an organist named Hendricks. They ended up having two sons and four daughters.

After Louis finished his schooling, he worked for a while with Philip Kusters in Goch. Then he went to Köln, Germany for further training with Peter Thelen. After these excursions, Louis resided in our parents' house and married (later on) Christina Theunissen. They had four sons and a daughter.

Johan Jr. was a great kid, always cheerful. He was a serious student and then became employed with the Rhine Railroad Company where he was held in high esteem. Due to his hard work and dedication, he became ill from a lingering cold. He died at the young age of twenty-seven in 1879. The authorities of the Rhine Railroad Company wanted, with the family's permission, to provide the gravestone for young Johan Jr.

Philip was a robust, well-built, young man. He wanted to become a baker and began working for his father's friend, H. Celis, in Venlo. He worked for 4 years but then became adventuresome like his brother Hubert and began traveling. In the beginning, he went from city to city but when he reached Diedenhofen (today it's Thionville, France), he stayed there a while since Hubert was there with the uhlanen, a German cavalry regiment. Hubert was still adjusting to military life and was a bit homesick. So Philip, to be near Hubert, got a job at a nearby bakery for a couple of months. Philip then worked in Kevelaer (which also happened to have a shrine that pilgrims frequented) for the Ophij Brothers and J. Verbunt. Philip finally returned to Venlo where he worked for Frans Bloem as chief baker. There he had an accident in which he fell with a sack of flour, resulting in a severe back injury. After a couple years in much pain, he died July 13, 1885.

II. My Youth

Now the story focuses on me, Hubert. I was born on January 28, 1855. After my birth, my mother became ill and was given the last rites. A good neighbor friend temporarily took me in. Fortunately, after fourteen days, I was able to return home because my mother had completely recovered. I became a strong and healthy child. At the age of four, I began going to school and initially performed and behaved very well. This good behavior didn't last long and as a result, the schoolmaster eventually had to complain to my mother about my conduct. Mother apologized to the old gentleman and asked him not to tell my father so I wouldn't get a drubbing. The following year, I had another instructor named Mr. Woring and things seemed to improve because he didn't tolerate any nonsense and believed in a firm smack to the behind of which I had my share.

So went by the first years of my childhood without any major mishaps except one. That was when one day after school, my friends and I decided to go to the town hall to jump down the stairs. We increased our jumps progressively from two steps to three and four. Even though my buddies had cautioned me, I was already off and running from the top of the staircase as they looked on. In an instant, I was at the bottom of the stairs and couldn't stand up. The boys quickly ran to a neighboring house for assistance and returned with a woman who helped them carry me, half seated and half lying, on a chair back home.

My mother, Regina, was as white as a sheet when she saw me carried in and immediately called the doctor. After a thorough exam, he concluded nothing was broken but that there were indications of ligament damage, something that could take a long time to heal. When no improvement was noticed 4 weeks later, I had to go to a specialist who had me lay in bed for 6 weeks where by traction my left leg was suspended and raised with a weight. But even after that, I was still unable to walk. So now the specialist fitted me with a left shoe which had a heavy weight under the sole and that enabled me, with the help of crutches, to slowly start walking again. That worked well and after 5 months, I was able to discard all these accoutrements and put the whole miserable incident behind me.

Even though I was a wild and energetic kid, I learned easily and made good progress in the parochial school I attended. The Very Reverend Alsters directed the school. At the age of 12, I was the youngest student to ever receive the First Holy Communion at my school. Most students had to be older than the age of 13 or 14. But as you shall see, my silly stunts didn't vanish altogether.

During the autumn of 1868, there were military exercises at the Dutch border. Boy, that was quite something when the soldiers arrived and our house had to quarter 16 soldiers. I soon made friends and received a helmet and a gun from them. The next day, I arrived late in school because I wanted to see the soldiers off for their morning march. The following day, the soldiers were going to a bivouac. The first encounter would take place near Straelen. To be able to see the action, I had to plan my strategy. Ms. Bel, our kind housemaid, promised to help me. The soldiers were already lined up in formation at the marketplace. I nonchalantly put my book satchel on my back and left home. Ms. Bel had filled my satchel with 6 sandwiches and a good piece of sausage and hid my books where no one could find them.

The soldiers marched forward singing in military cadence and I marched along with them singing as loud as I could. About 30 minutes before reaching Straelen, they came face to face with the enemy enactors and soon they were engaged in heavy fighting. As the infantry and cavalry immediately came to reinforce, you could hear the burst of gunfire. The artillery quickly followed and it wasn't long before one could hear cannons thundering throughout the sky. It was a deafening sound. But it was real neat. I loved it. The enemy was defeated and ran off the land. I always followed near the soldiers who were quartering at my house. The military exercises continued until 1 p.m. The soldiers were then able to take a brief rest before marching on in search of a place to camp for the night. It had been a beautiful clear day.

One officer who had already warned me a few times that I should leave told me again that it was time for me to return home. But it wasn't my intention to leave and I started to bargain with the officer to let me stay because it was my plan to become a soldier as soon as I had finished my schooling. The sergeant and a few soldiers who were quartered at my house saw my predicament and came to my rescue. They too tried to convince the officer that I was sincere and really wanted to learn about military life. The officer then laughed heartily and said to the sergeant, "Then you will have to take responsibility for your new recruit and give him the necessary things plus food." Then I opened my school satchel and showed that I had enough supplies for 3 days.

Some of the soldiers prepared the fire pits while other soldiers looked for enough wood and straw. Potatoes were provided and then the cooking began. What a treat, I was allowed to partake in the baking of the potatoes and the divvying up of the food. It was soon quite dark. The sentry fires were set up and a music ensemble was arranged. Music filled the air and the mood was just wonderful.

Suddenly, I, the young adventurer, thought of home, of Mother . . . and asked an N.C.O. "Where exactly are we?" since I wasn't familiar with the area. The N.C.O. then unfolded his map and said, "Here are the Rheurdterbergen (local hills) and there is Nieukerk as his finger pointed to the region. Tomorrow we return to Straelen." That made me feel better. It was dark and time to retire. The tents were set up and the fires glowed. At 9 p.m., taps were played. The music lingered over the camp. The fires gradually diminished. The guards walked along the tents and checked whether all the men had turned in or had found a place on a bundle of straw near the fire pit. I, the young recruit, was amongst the men near the fire pit.

Early in the morning, the guides reported activities of the enemy not far away. Quickly, all the equipment was gathered, some bread distributed and soon the loud boom of cannons came very close. The enemy came forward and as result, we had to retreat to Wankum, Broekhuizen and Straelen. There were some skirmishes off and on, but in the afternoon, it was quiet again and the soldiers returned to their quarters (in Straelen).

At the Dielen's was the table set with steaming hot soup as the soldiers cheerfully came in the house. I, the young recruit, decided to enter the house through the backdoor crouching low to the ground so that no one would see me. I thought it'd be better if I first explained to where I had been. But, boy oh boy, Father was ready with a big frown on his face and eager to give me, the young deserter, a deserved hit on the behind. Mother, however, interposed herself and pleaded with Father to give me, the young son, a chance to explain what had happened or where I had been. That gave me time to reach the room where the soldiers were without any corporal punishment. The good sergeant asked for a spoon and plate for me and I had to sit down with the soldiers. This broke the ice and I began telling about my adventures with the aid of the sergeant and the encouragement of the other soldiers. Even Father had to laugh now and then. It didn't mean that all was well and done because at 5 p.m., the Reverend Alsters showed up to complain about my absence at school. Lucky for me, my dad had just stepped out of the house. My mother's explanation gave the chaplain a good chuckle. The following morning the soldiers had to move on.

I returned to school after promising mother I would give school my best effort. But the consequence of my absence showed at testing time. I, who was one of the best students in the class, received an "unsatisfactory". So after a few days, the Reverend Alsters came for another house visit to conclude that I must leave school and would be better off being a butcher. So that was the high price to my spectacular experience with the army exercises.

School vacation had now begun. I had to go help Louis work the soil and keep the garden tilled and in shape. Louis was not at all pleased by the way I worked and complained to my parents that he'd rather do without me. I didn't like farming. It was not for me. I preferred to do the buying and selling for the butcher shops. So after borrowing some money from my mother, I went to the farmers and other butcher shops to buy the intestines and organs needed to make sausages. Autumn was approaching and that was usually the busy time for slaughtering. So, that would be a good venture for me.

The following Monday, I rose early. After a quick breakfast, I made my first stop at the local church to pray and then went on to Venlo to buy intestines (guts & entrails). Butchers Bootz, Bruil and Zweiszpfennig, who knew my father very well laughed when they saw me the young merchant negotiating. To encourage me, they accepted my low offer. I took out my pillowcase and stuffed it with 200 racks of guts and intentionally left some of the ware exposed outside the case so that it was also visible as to what I was selling. At noon, I was ready to return to Straelen but first I bought a ½ lb. of figs. Once I was outside the city, I ate the figs with my sandwiches as my midday meal.

In Straelen, a rack of gut was priced at 14 pfennig. I paid in Venlo only 4 pfennig. So I thought I had a good deal when I sold the gut for 14 pfennig in Straelen. I also sold some gut at good prices in Weiterbroek. By the time I arrived home, I had sold over 80 racks of gut. Everyone at home had a smirk on his or her face when they heard about my venture and I handed the money over to Mom. As Mother remarked with astonishment of my ability in repaying all the money, I noticed that although Louis didn't say anything, he was very jealous.

The following day, I again made my first priority to attend Holy Mass. Then I returned home for breakfast, put a couple of sandwiches in my bag and went about my business. By 4 p.m., I returned home having sold all of the racks I had with me. The next trip to Venlo involved the same routine. I arrived before noon at the Berg, which is over the bridge and to the right, but this time I didn't do as well. By evening, I had sold only 60 racks. It was satisfactory but less than expected.

My dad, Dielen Sr., was a butcher with the Crispin Pasch Company in Nieukerk. My dad's coworkers liked me a lot. They always butchered on Thursdays and had even invited me to assist them with the slaughtering. When I had told them about my new selling venture, they laughed heartily. On Thursday morning, I came with my dogcart so that in the evening, once the meat had sufficiently dried, I would be able to pile my portion of meat on the cart for my journey back home to Straelen. At home the next day, my father and I butchered two pigs and made sausages. My brother Louis was all the while working on our land with the help of a day laborer. I, your gut-merchant, was eager for Monday to come around again so I could repeat my visit to the local farmers for some more wheeling and dealing.

I had pretty much sold all of my merchandise when I returned home Monday night. On Tuesday, my father wanted to go to Venlo with me. Since he no longer walked such a distance, he sat in the dogcart, which was fine with me. But once we arrived in Venlo, my father wanted to walk on his own. He was a little embarrassed to admit to his old acquaintances and friends that I, his son, dealt in gut and intestines. Nonetheless, I went about my business of buying several hundred racks of gut. I stuffed them in sacks and piled them on the dogcart. Going home was not as easy as I thought. Father would constantly wave at people, stopping to say a word or tell a story. It was pretty dark and late when we finally got home. I now had enough to sell for a couple of weeks.

During this time of the year, the farmers would slaughter their animals. I decided to offer my help in the carving and salting of the meats. I got paid only 5 groschen. It was a miserable wage for two days work. It is understandable that I would rather be selling guts than being a butcher, since a gut salesman got paid a whole lot more to sell a 10 rack of gut than to carve and prepare a whole pig.

So the winter came and during the long nights, I would find some extra work at carpenter H. Janszen's or even at the shoe shop of the cobbler Euprathen Thomes. Although the butchering season had come to an end and didn't pay well, I learned a lot. I could butcher and carve a whole pig all by myself. That was pretty good considering my age. Of course, I could have been of good use as an extra helper on my parents' field but I would do anything to avoid that kind of work. So I found an excuse that would suffice. I was going to buy young goats. It was kind of a low job but that was all I could think of. My mother agreed to this arrangement provided I would earn enough in a day to pay for the day laborer in the field. So I had to earn at least 6 groschen a day. That was fair enough. I bought some goats for 12 to 15 pfennigs. I'd slaughter them, skin them and then hang the skins to dry. The meat would be sold to people who had dogs.

One potential buyer for the goatskins was a little Jewish fella who complimented me on how well the skins were prepared. I wasn't sure what price I should ask, so I asked him what he thought the price ought to be. He offered a bid of 3 groschen. I replied, "6 groschen." He scratched his head and made a counter offer, "4½" I didn't agree. I'd rather wait. I had after all, 260 skins to sell. Another Jewish merchant stopped by but I didn't agree to his offer either. I was confident I would get at least 5 groschen.

My mother started to worry that I wouldn't get what I wanted and end up stuck with the goatskins. My father joined the discussion and said, "Get rid of those smelly things, get them out of the house." No one came knocking on the door and even I started to worry. Eventually another merchant stopped by and I was able to sell the entire batch for 5 groschen per goatskin. Mom really beamed when I handed her those beautiful thalers.

A couple of months later—a butter, eggs and milk dealer, named Kusters, came to town and offered me a job. Although Mother was strongly against the idea, I thought I'd give it a try. I would stay with the new job only as long as it didn't interfere with the other things my mother asked me to do. I would go to the farmers in my dogcart to collect butter and eggs. Mr. Kusters, the merchant, would rendezvous at the inn on the main road with his horse and buggy. I would go home when it got dark, while Mr. Kusters stayed to play cards and drink. He was really addicted to card playing. I did this Monday through Friday for 5 thalers a week. This routine repeated itself week after week.

One Monday morning, I had to be in Pont at 7 a.m. with my dogcart. My brother Johan, who happened to be at home, wanted to go with me to see what it was like. We both had some well-prepared sandwiches in our bag and off we went. By midday, we had finished our job and one of the farmers offered us coffee and eggs. After this meal, we went to Walbeck and again collected butter and eggs. At 4 p.m., we turned around toward Straelen. We made a short stop at the Inn, *The Three Kings.* In this area, I felt quite familiar and started to round up more butter and eggs at the local farms. Mr. Kusters also visited some farms to gather butter and eggs. When Johan and I met up with Mr. Kusters at *The Three Kings* Inn, Kusters was again playing cards with two friends. He asked us to make one more round at the farms to collect more butter while he played cards. Johan and I worked until 10 p.m. and returned to the Inn. Kusters was still playing cards and was very intoxicated. Brother Johan was quite disturbed and thought we should just leave and go home. I didn't think that was right and wanted to wait for Mr. Kusters. We decided to wait at one of the farms. We asked for a glass of milk and ate our sandwiches. The wife of the farm owner was a friend of our mother. She suggested we go home since our mother would surely be worried. So we left, but first we checked again at *The Three Kings.* The innkeeper was ready to go to bed. We were informed that Kusters and his two buddies had left stone drunk with a horse and cart. So we took our dogcart and drove off in the direction of Straelen. Johan sat in the back of the cart and I was the driver. The dogs, which by now were also eager to get home, ran in full gallop straight to our courtyard.

Father and Mother had been so worried about us that they were praying the rosary. Johan and I explained what happened as Johan concluded, "Never again." Father, who was usually not too uptight said, "That's the last time," as Mother cried. I thought to myself, oh darn! Now I'll have to return to working in the field.

The next morning, I didn't come downstairs until 9 a.m. Now Johan and I had to decide what to do with all the butter and eggs that were in the dogcart. We thought it best to just wait for word from Mr. Kusters.

Two days passed before Johan and I found out something. There was a rumor that Kusters had left town. A few days later, a tall, skinny, and very pale woman arrived at our house. It was Kusters' wife. She was sobbing very hard and could barely speak. She handed a letter to our father to read. Her husband had written to her that he had sold everything and had left for Antwerpen, Belgium. (Another emigrant.)

Father escorted the woman to the town hall police. The police promised to investigate. The woman with the help of my father sold the butter and the eggs that I had hauled home. I received 6 Deutsche marks for my work and the poor miserable woman returned to Wachtendonk. But now what was I going to do? Return to working in the field? That was not something I felt like doing; nor was much butchering done this time of the year. So that left being a merchant my only remaining option.

I was back to my old routine the following morning. That is, going to Mass, eating breakfast, and then to the garden or working in the field. Although I had put much effort into thinking about what other work I could do, continuing to sell was temporarily out of the question. Any excuse to shirk my duty in the field, was done. Many a time, I would simply leave my tools on the land as I went with my younger brother Philip and his buddies to look for bird nests or catch salamanders.

There was talk that in the garden of our pastor, a scarecrow was mounted dressed up like a policeman. The complete uniform even included a *Pickelhaube,* or "typical German pointed helmet." The pastor did this because in the past all of his beautiful cherries had been stolen from his trees. I pretended that the birds as usual had eaten them. The naive pastor even thought he could succeed in concealing the knowledge that the policeman was just a scarecrow.

Four of us young fellows got together at 11 p.m. Saturday on the outskirts of the pastor's garden to sneak into the garden from the back. But first we had to reassure ourselves that it was the scarecrow on duty and not the real officer, Amwech. Bernard von Schützing, the son of the burgomaster, volunteered to sneak closer and investigate. Once he was sure the scarecrow was on duty, he grabbed the thing and danced around with it.

The garden of the rectory was separated from the house by a stone wall about two and a half meters high. This wall gave us the privacy to pick and eat the cherries to our hearts content. We even stuffed some cherries in our pockets to savor the next day. But now came the question of what to do with the scarecrow. We quickly came to a solution. We grabbed two long beanstalk support sticks and pushed them through the back of the coat of the scarecrow and placed him on top of the wall near the door. Suddenly, we heard a noise. It was the night watchman coming our way. The scarecrow was quickly retrieved from the wall and placed in front of the garden entrance. We hid nearby under the bushes. The night watchman snickered when he saw the scarecrow and thought it quite funny that the pastor had such an ingenious idea. The watchman then walked on. We were lucky that we had completed our cherry picking five minutes earlier because otherwise the watchman would have encountered us. We again placed the scarecrow on top of the wall and then went home unnoticed by the watchman because we knew where he was.

The following morning was a Sunday and the countryfolks from the nearby hamlets came to attend the first Mass. It wasn't long before everyone had noticed that the scarecrow had been placed on the reverend's orchard wall. The scarecrow must have been intended to portray Amwech because he was the only one in the city of Straelen who wore a *Pickelhaube.* I had attended the 8 a.m. Mass and afterwards had to see the spectacle myself. There were about a hundred people gathered around the scarecrow. Some thought the displacement a disgrace, others a good laugh. As soon as I mentioned that the whole thing was a good joke since the birds would have eaten the pastor's cherries anyways, a whole group of women immediately surrounded me and yelled into my ears, "Don't you have any respect for our venerable pastor?" "Oh, go to hell," I said and then quickly got out of there and went home.

As usual, I went back to church to attend the High Mass after the early morning Mass at 8 a.m. When I got home, my mother was waiting for me with a deep frown on her face. She told me that the mayor and the chief of police wanted to speak to me in the town hall.

When I arrived at the town hall, I was met by the policeman Amwech and two other officers. The burgomaster sat at a table. The burgomaster asked whether I knew anything about the scandal and theft. I asked the burgomaster whether I could explain everything candidly. He said, "It was not only allowed, it was expected."

I told him I heard the news after the Holy Mass of 8 a.m. and had to see the spectacle for myself. It was then that I said it was quite a neat joke. The women around me asked whether I didn't have any self-respect. When I replied that the cherries only went to the birds, they started to yell and scream and wanted to attack me. I yelled back and made sure I got out of there before things got worse.

The policeman who still remembered me from the military exercises could not hold his laugh any longer. Now the burgomaster wanted to know where I was last night. I replied that I, as usual, had been to Herongen, Niederdorf and Louisenburg with the meat cart and had returned at 8 p.m. After dinner, I had gone to *The Bocksteeger* to play some cards and then went to bed around 10:30 p.m. My brother Louis even awoke when I returned home and asked me what time it was. I replied, "Half past ten." The policeman interjected, "That's right, I asked his brother Louis and he said the same thing." To which I continued, "And that's why I think, Mr. burgomaster that you're on the wrong track. Besides, is this case really all that important? After all, it's just a youthful prank that was probably done by your own son anyways." I gave a wink to the policeman who took it from there. "I'd like to make a remark Mr. burgomaster, that if your own son would've done it, you would've probably found it amusing and the whole town would have laughed about it. And if you had to appear in the city of Kleve at the court of justice, then those gentlemen also would have laughed it off. Let us just forget the whole thing. It's still a theft, but it's my understanding that no one has filed a claim and if the Pastor shows up at your office, you can tell him to look into it himself." So that settled that. On my way out, I asked Mr. Amwech whether he was going to sit on the pastor's wall again. Under roaring laughter, we exited the town hall to the great hilarity of the awaiting public.

Mother was still at home waiting in great anguish but Father stood there with a smirk on his face. He cautioned me to watch my behavior because the next time it might not end so smoothly. He knew that I knew far more about this story than I let on. I went upstairs and returned with a handful of juicy beautiful cherries. My parents, however, declined to eat any so as not to become entangled in this boyish adventure. So that was one of the several pranks that occurred that hot summer.

As time passed, I engaged in a new enterprise of buying and selling puppies. On a Monday, I was again dealing with the farmers but didn't do so well on my first day. I was only able to buy 6 puppies since I only had enough money for 2 to 2½ groschen per puppy. So I offset that by selecting the biggest ones. At the next address, I spent 5 groschen. By the following day at 4 p.m., I had my 30–groschen's worth of puppies (about 14) and returned home. Since I didn't want to carry the puppies in a basket, I instead bound the paws of the little animals and slid a long pole between their legs (before I placed the pole behind my neck on my shoulders). While I was in a narrow street, one of the puppies became untied. So I put the pole with all of the puppies on the ground and quickly snatched up the runaway. But, oh my! When I had returned, all hell had broken loose because now half of the others were on the run. I didn't expect puppies to be able to run off with their paws tied. It was now dusk and I was able to retrieve all but a couple of the runaways.

The next morning a farm boy came to our home to bring the remaining puppies I wasn't able to find. My family wanted to know what this was all about and I was happy to explain. Everyone thought it was quite a story and laughed wholeheartedly. I was able to sell my puppies easily and profit a good penny. But I also learned I shouldn't engage in my enterprise without a dogcart or a basket on my back. I also knew my father didn't want to give the cart so that was the end of the puppy business.

During this time, my brother Louis was working in Köln for a cattle merchant named Lormans. Louis was learning how to make sausages in Köln and Goch. Lormans was born in Geldern and often traveled between Geldern and Köln. Lormans also knew my father very well. Louis was no longer doing much butcher work and so this provided me an opportunity to fill the void in Geldern and work as a butcher, even if it was only half a cow a week in addition to one calf and one or two pigs. When I had spare time, I would work in the family garden. I gradually began to enjoy that kind of work.

The gut market was beginning to pick up again. Winter slowly came upon us. But another problem came about. A dog with rabies had been spotted in town and was immediately shot. To prevent the spread of rabies, all dogs had to be chained and kept on their premises. This meant that on Saturday, we couldn't use our dogcart and thus had to personally carry the meat to Herongen. Father had to take the straw basket out of the attic (see the glossary at the end of the book for the word *kiep*). Everyone had to take turns carrying the basket. The straw basket was of course carried on the back. We decided beforehand how far each person would carry the basket. We mapped out the route to Sant, Broekhuizen, Herongen, Niederdorf, Louisenburg and back. Boy, it was quite an undertaking. In addition to that, I had to assist the farmers with butchering for only 5 Groschen. I became fed up with it all.

III. The Journeyman Period

I was continuously planning for the day when I would leave home. Then the big day arrived. I departed on March 29, 1872 on a youth trek. When I said goodbye to my mother, she was terribly sad. The tears rolled down her cheeks. When she held my hand, she pressed a rosary in my palm and said, "Hubert, dear, don't forget The Good Lord."

Off I went on the long haul. I found work in Kempen with a Mr. Gelts. The job didn't teach me much and I didn't feel like hanging around, so I moved on to Krefeld and stayed with a Mr. Winterschrei. From there, I went to Mr. Haffner. Neither place showed any promise and the treatment I received was worse than what a dog would get. So I moved on to Neuss and stopped at a butcher shop. There, however, the owner was the whole week on the road. His master butcher and the lady of the house, a rich farmer's daughter, began to enjoy the owner's absence and began an affair. Once I realized that, I decided that wasn't a wholesome atmosphere and packed my knapsack to move on. I arrived in Düsseldorf. There, I wasn't able find a job. So on to Duisburg to a Mr. Pasch, someone I knew from my days at home. I was curious as to how he lived but it was a real disappointment. His wife was always drunk. She sure loved peppermint liquor. I wrote a letter home to explain where I was. By this time, I had been gone from home 14 days. My parents replied right away and advised me to leave that house immediately. I stayed a little longer but then decided to follow my parents' advice and move on. First I went to Köln, but I couldn't find a job there so I went to Bonn. There I found a job with

Map of Europe Around 1875

a Mr. Karl Nap on Neugasse (New Side Street). They were real sweet people and Karl was a pretty good butcher but they were just looking for a helper in their shop. I felt my skills were higher than that. So after a couple of weeks, I strapped on my knapsack again. The knapsack was called a *Berliner* and was the traditional travel bag for any journeyman. It was hung over the shoulder with a green strap. (See Glossary)

Being a member of the *Kolping Journeyman Society,* one would first look for the Kolping lodge when arriving in a new city. Your suitcase would then be immediately forwarded from your last address to your current location. The next morning, four of us associates (a blacksmith, a baker and two butchers) left the inn and journeyed along the Rhine. Soon we were singing out loud, *Ein Straüschen am Hut, den Stab in der Hand,* or "a posy in the hat, and a staff in the hand." The road went first to Bad Godesberg, Königswinter and then to the Zevengebergte (local hills). We continued on to Brohldal, Ahrdal, Remagen, and then to Koblenz. In Koblenz, I stopped by all of the butcher shops seeking work, but nothing was available. They did offer me, though, some sausages and that was great. I then passed through Ems, Nassau, Wiesbaden, Frankfurt and on to Mainz. I stayed longer in those larger cities to search for work and to find some food, actually to beg, because I was completely out of money. In the beginning, that is hard to face. But as poverty teaches you prayer, hunger teaches you begging. Of course, we didn't want to miss any of the beautiful sights, too many to mention. I then went to Worms, Ludwigshafen and Mannheim. The competition for work was in full swing. Large groups of young men would come daily from Strassburg (today it's Strasbourg, France) to look for work. So my chances of finding a job were nil. So I decided to leave Mannheim and move on toward Heidelberg.

Finally, I was able to find some work (in Mosbach) with a man named Felix Letzkus. He was very weak and in poor health but his family was very kind. A couple of days before I arrived in Mosbach, I had passed through Neckarelz. There, I had to beg here and there for a piece of bread because there were no butcher shops where I could ask for work. At the last house, somewhat away from the road, stood a little old hag at the door. I also asked her whether she had any spare bread. She grinned and nodded and invited me to come in. She told me that she understood my situation. She too had to beg for bread now and then. Inside the small house near the front door was a kitchen and a large fireplace where an open fire was burning. Then there was a small room with two 50 centimeters square windows. Below the windows was a small narrow chest on four legs. The chest was used as a bench. In front of the chest was an old table with 2 wooden chairs. The little old woman felt quite honored by my visit. She lifted the lid of the chest and took out a piece of bread, a piece of cheese and a pocketknife and bid me to eat. Boy, did it taste good. I was starved. The room had a dirt floor with the exception in the corner where two planks of wood lay. The little old woman then lifted one of the planks to pull out a jug and brought out a small drinking glass from the chest. "This is the finest cider, please take some." I didn't need further encouragement.

While I ate all I could, the little woman talked and talked and asked many questions. At the moment I wanted to leave, I reached out to shake her hand and thank her for her hospitality but she had one more offer. She pushed aside a cloth that hung in another corner, behind the cloth stood a chest. It was filled with straw and some rags. It was her bed. Underneath the rags was a small purse with some coins in it. She gave me 2 baksen and 6 kreuzers (German money). While she made the sign of the cross, I waved to her good-bye and I left smiling. The little woman stood in the doorway with her beautiful black cat in her arms. "Be good and careful young fellow," she yelled out as I went on my way.

When I finally arrived at butcher Felix, as he was called in Mosbach, my clothes had worn down to rags. I was soon the favorite of Felix's wife and I got along just fine with their only daughter, Liesel. When butcher Felix went out for recreation, I went along. I had written home for my suitcase, which arrived promptly. Once I had my better clothes from my suitcase, Liesel came along too. Even though Liesel was 6 years older than I, she was obviously in love with me. We strolled along the wine orchards. One Sunday afternoon, we promenaded all the way to Neckarelz. In the distance, I saw the kind little old woman with the beautiful black cat in her arms. She stood on the front doorstep of her little old house. Liesel whispered to me that the woman was a witch. Butcher Felix and I had to heartily laugh about all the fantastic stories Liesel was telling us. As I was telling butcher Felix and Liesel the experience I had with the poor little woman, we reached the little house. I wanted to repay the woman for the kindness she had bestowed on me 4 weeks earlier. So I asked butcher Felix whether he could lend me a good guilder. He handed it to me without a second thought. Liesel became frightened and asked what I was going to do with the guilder. Butcher Felix thought my idea was grand but cautioned me about becoming too familiar with the little woman even though he didn't really believe in witches.

Liesel, however, clinched her father's arm and was pale with fear. She was afraid of our mere encounter with the witch. At first, the little woman didn't recognize me. But when I offered her the guilder, she started to cry with joy. While continuously making the sign of the cross, she told me that she had said many prayers for me. Since we were in Protestant territory, her gesturing was conspicuous. Liesel felt uncomfortable and pulled us on our way.

The relaxed atmosphere of our promenade was gone. We stopped at the first best winery and talked about the encounter with *the witch* in a laughing way sustained by flowing wine. To my surprise, I found the whole village knew that a young journeyman had eaten a meal with the witch. When I told the group I was the poor fellow and the little woman had been exceedingly nice to me, they were all confused. A fine young gentleman had accepted bread and wine from a *witch?* It didn't take very long before all the villagers of Mosbach knew that *Robert,* as I was called in Mosbach, of butcher Felix had befriended the witch of Neckarelz and that he, like the witch, was a *crosshead,* or "Catholic". To anyone who asked, I would retell my experience with *the witch* and would add how surprised I was that the villagers of Mosbach were so stupid as to believe in witch stories. Such an opinion didn't make me very popular and it didn't take me very long to realize that my good days in Mosbach were over. I picked up my *Berliner,* or "knapsack," one early morning and journeyed on.

My journey continued on to Kehl and Strassburg. Although several homes on the *Steinstrasze,* "Stone Street," became uninhabitable due to the bombardment in the war of 1870, the city of Strassburg was not too much impaired. The Münster cathedral was not much damaged either. Although its glass windows were covered with boards, the renowned clock still worked like it always did.

I wasn't able to find any work except at a butcher shop that belonged to the Army. Another journeyman and I had asked for a job there and we were able to begin immediately. We were offered 6 Deutsche marks a day, payday being every 5th day. We left after the third payday. That job was impossible. One had to sleep on straw and it was your own responsibility where your next meal came from. That was too much. We were too young for that.

So my journeyman buddy and I journeyed on through the Black Forest via Oberkirch, Oppenau, Freudenstadt and so on, slowly to Schaffhausen and Basel in Switzerland. We, however, soon encountered the Swiss police. Since we didn't have the correct passport, we were returned to the German side. We then went in the direction of Donaueschingen and Urach. We wandered through the mountains but had the bad luck to lose our directions twice. We had to stay overnight high in the mountains. In Urach, an old quaint village nestled in the mountains, we found work.

Although my clothes had suffered from all the wandering, they were adequate for the job I found. I was working for a Frohmann Röslewirt, who ran a boarding house with an annex butcher shop. The guests there looked more like robbers than tourists. Urach was supposedly known for tourism. The owner of the boardinghouse was drunk every night and played cards with his guests. The card playing usually ended in a brawl.

One evening got pretty bad, the boss had cheated with the cards and someone gave him a good beating. Half of the inventory was destroyed and he bled like an ox. The next day, almost no one showed up. After I was through with my routine, the boss asked me to come over. He wanted to play cards with me but I told him I didn't play cards. He drank one shot of gin after another. I was sitting at a corner table reading a magazine and smoking a pipe. All of a sudden out of nowhere, he grabbed me. I dove under the table but he continued to kick me so that I could barely get out of the way. Then I got ahold of a chair leg that was still lying around from last night. I now came out from under the table but Röslewirt in grabbing ahold of me tore my coat. That did it, now I pommeled him. Soon his wife showed up and attacked me from behind. I yelled for help and luckily a night watchman happened to be walking by and came to my rescue. He had to climb through an open window because the door was locked. Once I was outside, I was told the police had been watching Frohmann for a long time. That was encouraging news in case I needed it for my defense.

The night watchman and a husky policeman escorted me to another lodge where the innkeeper was ordered to provide me with a good meal and bed. When I told the innkeeper about my incident with Röslewirt, he told me Röslewirt had a lousy reputation and I was lucky to escape with not more serious injuries.

At 9 a.m., I had to appear before the chief of police and tell my entire story. The police then had Frohmann picked up. Frohmann, of course, denied everything. But the night watch, who had witnessed the event, confirmed my account of the story.

The police officer took me to a tailor to get my entire suit repaired. Even my shoes were resoled. One of the fellows who worked in the tailor shop loaned me one of his suits so I could dine at the local inn. Frohmann and his wife also entered the inn and wanted to talk to me on the side. The innkeeper didn't allow it, saying he was obeying the chief of police's orders. The wife of Frohmann persisted in her scheme by trying to shove a small bag of money into my hands while asking me to get out of the area. When the innkeeper saw this, he grabbed the woman by the arm and pushed her out of the door. Now I had a chance to eat and not sparingly. I ate an amount equivalent to two days worth. At 4 p.m., my suit was ready. The shoemaker had returned my shoes and now I was dressed well enough to return to the chief of police. I had to retell the whole story again; now with the addition that Mr. Frohmann attempted to bribe me with money. The clerk prepared the police report and the chief of police asked that I sign my name. Mr. Frohmann was then summoned and my deposition was read aloud. Frohmann was instructed to cosign the report but he refused. That meant arrest for Frohmann; whereas I was escorted back to the local inn and after another good meal, went to bed.

The following morning, a police officer showed up at my breakfast table with 5 guilders for a day's work and 8 guilders compensation for the beating I had endured at Frohmann's. The police officer wanted to know whether this was satisfactory to me. Of course it was. The police chief also paid for the room and board at the inn.

The old town Urach is situated in a bowl and could only be reached from one side unless you wanted to go over the mountain ridge. Naturally, there was no railroad line either. The police officer was assigned to take me to the nearest railroad station. Traveling by road would take about 8 hours, but we could make it in 3 hours if we hiked over the mountain ridge. So we decided to go over the mountain ridge.

The hike was exhausting but the scenery was beautiful. During our trip, the calm fifty-year-old police officer told me how dangerous the *Röslewirt* really was. But now, the police really got Mr. Frohmann Röslewirt. The police officer told me to be sure to send a note to the police chief after I had found a new job. As soon as we arrived at the small railroad station, we ate and drank well (at the local café). Then the police officer bought me a train ticket to Ulm. We warmly said good-bye and that concluded that episode in my odyssey.

Arriving in Ulm, I went first to *The Golden Hand* inn. From there I began my job search. All the while, I enjoyed the beauty of the city. The historic city was near the Danube River and had a large cathedral, which housed the largest organ in the world. Everyday, between noon and 1 p.m., an organ performance was given especially for the tourists. Since after 3 days I didn't succeed in finding work, I eagerly accepted a rafter's offer to go along to Regensburg. He was a pleasant fellow who really enjoyed hearing all of my stories about home, parents and travel experiences. We soon became the best of friends. The riverbanks of the Danube were extremely beautiful. The quaint small towns with their old castles were truly a romantic sight. We would anchor here and there at my request. In Günzburg and Dillingen we did some real sightseeing. We also landed at Donauwörth where a church festival was taking place. We didn't want to miss that. Donauwörth is a quaint town where a small stream, the Wörnitz, flows into the Danube. After some food and a good glass of wine, we walked toward the church festival site. At the marketplace, where the festivities took place, were some wooden planks that constituted the dance floor. Under a few tall old linden trees were seated 4 musicians who played dance music. Although I knew the music well, the way they were dancing was unfamiliar to me. The rafter, who felt right at home, promptly asked a girl from Köln to dance with me. My rafter buddy insisted I dance with her, since I had been boasting about my dancing abilities. Once the music started up again, the girl and I began waltzing to the music. We didn't talk much and the spectators attentively watched our dancing. The rafter was having a good time, so we stayed at the festival for quite a few hours. When the sun started to go down, the music ended and the young woman from Köln strolled with us to our raft. After a warm good-bye with many kisses, it was time for the rafter and me to board the raft and drift downstream. Once darkness set in, we set anchor again. We ate something and prepared for the night. We improvised a hut from some boards and used straw as a mattress. We rose early the following morning and set out on our final stretch to Regensburg.

Arriving in Regensburg, the rafter and I had to combine several wooden rafts to form one large raft. It took about 4 days to build. The rafter, who by now knew me quite well, asked me whether I wanted to continue on as a workman and later on as a supervisor. But that didn't appeal to me. So in good friendship, I said good-bye and went on my way.

After crisscrossing Regensburg for 3 days without finding work, I decided to journey on to Munich via Augsburg. In Munich, I didn't have any better luck in finding work and the police noticed on the second day that I was begging for food. I had to run to avoid being arrested. When I arrived at Innsbruck, Tirol, there were no jobs available there either. After 2 days, I was already begging for food. The police wanted to take a look at my passport. I didn't have one, so I was told to return to Germany. I kept a low profile as I trekked along the Austrian border back to Regensburg and then on to Nuremberg. When I arrived in Nuremberg, the great meat and beer revolt had begun. The prospect of my finding work there was nonexistent and so my best bet was to quickly move on. After a few days, I arrived at Fürth.

I finally found work at the *Three Kings* hotel. In the mornings, I'd prepare the meat for the hotel kitchen. Then I'd go with the chief butcher to the meat auction hall where we sold (our surplus) meat. Then I had to return to the hotel kitchen to assist with the cooking, braising of meats, meat carving and if needed, butchering. When that was all done, I was allowed to assist in waiting tables. Boy, that was right up my alley. I was able to make a pretty penny.

The people in Bavaria have a different diet and eating routine than do the people in Straelen. In Bavaria, we'd get up at 5 a.m. and work until 7 a.m. Then we'd have a roll and a large cup of coffee.

Then after working till 9 or 10 a.m., it was time for breakfast. Black bread wasn't familiar here. A large loaf of white bread weighing 8–10 lbs. was passed around. One would hold it to the front of his chest in order to cut a good chunk of bread off of it. He'd then transfer it to the next person until everyone had his share. Meat sausage and cold cuts were also plentiful. The apprentice would cater the tankards of beer that we used to refill our steins. So, since at 10 a.m. the beer taprooms were all busy, people would go first to the butcher for a piece of sausage before they went to the taprooms for a beer.

Noontime dinners were awful. They usually consisted of only peeled potatoes with buttermilk. At 4 p.m., we had another cup of coffee with a hard roll. Then at night, we again had bread with beer.

My supervisor was jealous that I was making so much money waiting tables and after awhile, I thought it better to don my knapsack and be on my way.

Luckily, 2 days into my journey, I was again able to find a job in Bayreuth. I was working for a fellow named Paulus Knör, who owned a fine sausage factory. Within 2 weeks, I was working in the store next to the owner's wife. It was a large business with 8 butchers and 2 sales girls. The owner no longer did any heavy work but still took care of the purchasing. Fritz was the supervisor who was in charge of the factory. Second in command was a good-hearted, rough looking guy named Leonhard Klötzer. Unfortunately, he was drunk everyday. I was third in rank. Next in rank were 2 volunteers. They were rich farmer sons who were assisting the business in order to learn and so forth.

The pay was low. The supervisor made only 5 thalers per week. Klötzer made only 3 thalers per week and I made only 2 thalers per week, which consisted of 12–14 hour workdays. By helping Mrs. Knör on Sundays, I was able to make an additional thaler. The factory butchered about 2–3 bulls and 30 pigs a week. Being third in rank, I had to go with my supervisor twice a week to a cellar that had been cut into the rock of a hillside on the outskirts of the city. We did the salting of the meats there. Then we would stack the cart with hams, sides of bacon and so forth before we'd call the carter who'd pull the cart and we'd push or walk behind it. The supervisor seemed to enjoy my company but he was more at ease with the carter of whom he was well-acquainted.

On the way to the cellar at the end of a street was a brewpub named *Hackenbräu.* There we would each have 2 mugs of beer before going on our way. The supervisor always insisted on paying for our drinks. That was fine with me. Once we arrived at the cellar, we swung the door open and the cart was pulled inside. Fritz, the master butcher and supervisor, let me do the salting. The following morning after my first day at the cellar, in the owner's presence, Fritz complimented me on my work.

This work routine continued for a couple of weeks until one day I started to count the meats on the cart after the salting. There were only 4 pieces, 5 pieces short! I didn't say anything. When we returned on Friday to the cellar and afterwards stopped at the *Hackenbräu* for a beer, I decided to step outside and take a look at the cart. Lo and behold, the wife of the owner of the brewpub stood there with 2 large hams in her hand. When I asked her what she was doing, she replied, "I'm admiring how excellently cut are these pieces of ham." She then placed them back on the cart. After the salting procedure in the cellar, there were again 3 pieces less than when we left the Paulus Knör factory. Again, I didn't say anything.

On Sunday, after the supervisor Fritz had left on the first early morning train to visit his fiancée and we in the store had finished our work, I asked for a moment with Mr. and Mrs. Knör. After they agreed to my confidentiality, I explained how they were being robbed. Mrs. Knör immediately began crying and said, "No wonder why we aren't making any profit even though we have such a spectacular business." At 4 p.m. that same day, Mr. Knör had to go to the market in Leipzig to buy livestock for slaughtering.

Fritz, the supervisor, was supposed to be at work 7 a.m. Monday, but didn't show up until 9 a.m. One of the volunteers, who often worked with Fritz, also came late. The remaining workers were working as fast as they could but weren't able to keep up. The two latecomers usually handled the large meat cutting machines with 8 knives. Nothing had been done yet in that area. There were still 600 lbs. of meat to be cut.

Without sleep and still half drunk, the two latecomers started to work. But, oh my! We all of a sudden heard a terrible scream. The volunteer who worked with Fritz had left the cutting machine unsecured and when Fritz came closer, the knives came down on Fritz's arm and hand. A part of his arm and two fingers were laying on the chopping block. Fritz's coworker, who was the volunteer, was horrified at the sight. He took off along with all the other workers. The second in command, Leonard Klötzer, after downing a dram of whiskey, was the only worker who assisted Fritz who had fainted. I ran to Mrs. Knör and told her what had happened. One of the sales girls immediately ran for the doctor who lived nearby and soon Fritz was on his way to the hospital. When I returned to work, Leonard had pretty much cleaned everything up and we continued with our duties.

At 6 p.m., earlier than expected, the owner had arrived home from his trip to Leipzig. When he heard from his wife what had happened to Fritz, he replied, "He must have had long fingers." At least, that's what one of the apprentices had heard him say.

After a few days, Fritz returned to the company. Fritz at first received a chilly reception from Mr. Knör but then the owner decided to allow Fritz to continue on the job as supervisor if he could handle it.

After thinking it over, Mr. Knör decided to report the theft to the police. A plan was devised with the police to trap the thieves. I didn't go to the cellar anymore. Remarkably, the detectives, who were sent out on Friday night, caught the thieves. The thieves were Fritz, another coworker, and the wife of the owner of the *Hackenbräu*. At Knör's house, everyone was very depressed. Mrs. Knör cried bitter tears and said, "Only God knows exactly how much has been stolen from us all these years. We were always wondering why we were unable to make a profit while working so hard!"

But in the factory, there was immediate talk about an informant amongst them. Had not the owner mentioned that Fritz had long fingers? (An idiom for thief.)

Later at night, Leonard Klötzer, being half drunk, brought up the subject again. But when I noticed that his suspicions were directed at me, I had to ask, "You're not referring to me, Leonard, are you?" He replied only a Prussian would've squealed. Since I was the only Prussian, I knew he was talking about me. I then realized it was in my best interest to leave. So that same evening I told the boss of my intentions. His answer was short and to the point, "I'll throw the whole bunch out." His wife begged me to stay but Mr. Knör thought that for my own safety it'd be best that I leave too. So I packed up my *Berliner*, or "knapsack," and put the rest in my suitcase, which Mr. Knör promised to mail to me later.

I gave a curt farewell to my colleagues and off I went. I was back to my *wanderlust* again.

At first I tried to enter Bohemia by hiking to a small border town that looked like Venlo, being composed mostly of farm fields and all, but it was just my luck to be spotted by the police on the second day. They delivered me right back across the border. So I journeyed toward Saxony through Hof, Plauen, Zwichau, and Chemnitz. It was in this region I passed by the *Pflauischen Grounds.* That is where the Royal Saxony Silver Mines are located. From there, I continued on to Dresden.

Dresden is a beautiful city with parks, plazas and large bridges over the Elbe. In the small coffee shops around the plazas, one could get a cup of coffee with a sweet roll for only 10 pfennigs. There were also many men and women carrying a yoke on their shoulders that had on one end a basket with long rolls and on the other end a kettle full of hot Vienna sausages. You heard them all day yelling, "They are warm now," to which the naughty boys would add, "and stink too." For the price of only 2 groschen, one could get a roll and sausage with as much mustard as desired. It was a real bargain.

After a couple days of being unsuccessful in finding work, I continued on along the Elbe to Meissen. During these past few weeks, a fellow named Karl Stroh accompanied me. He was from Florsheim, near Mainz. I still have a photograph of us that was taken at Dresden. Meissen is an old town along the Elbe with an old citadel, The Albrechtsburg, located on top of a hill. Albrechtsburg was used as the courthouse and jail.

My buddy Karl and I found a place to stay with the Kretschmar sisters, Rosengasse (tr., Rose Lane) 329, who were willing to rent to us. The family, which was pensioned, consisted of two sisters and a brother. The Kretschmar sisters, named Elise and Emilie, were both around 30 years old. Their brother Helmuth, who also lived there, was 35 years old and slightly retarded. Helmuth was anxious to tell my partner and me they had a childless uncle in Berlin who owned a large meat market. Elise was leaving the following month to Berlin to be adopted by her uncle. The sisters gladly offered me the address of their uncle in Berlin in exchange for an address of an uncle of mine I had jokingly mentioned because Elise wanted to meet me again. Unfortunately, my uncle in Berlin was a fictitious persona. Lucky for me, Karl quickly knew how to change the subject. He brought into the house a blind harmonica player and a couple of young girls from the neighborhood. We then had a grand old time of laughing, drinking, singing and eating until midnight.

After a good night of sleep, Karl and I asked for a cup of coffee but were instead treated to a complete breakfast. When we wanted to compensate the sisters with some money, they declined our offer. Elise smiled and said we could repay her later when we were in Berlin. We shook hands and sealed the agreement with a kiss on the cheek. The sisters weren't exactly pretty girls but they were very sweet. After that, we went by some butcher stores to replenish our sausage supply before we were again on our way.

Karl and I journeyed toward Döbeln, where we arrived at night. The walk didn't seem very long because we were so busy talking about all the fun we had had the previous evening. Although the local inn was pretty crowded, we showed our identification card at the entrance and paid 2½ groschen for our lodging. We then ate a piece of bread and went to our respective sleeping cubicles. We immediately conked out.

Suddenly, I felt an ice-cold hand on my bare arm and I was asked to get up. I told the person I had paid my lodging and wished to be left alone but to no avail. Because when I finally came to my senses, I was looking up into the eyes of three policemen who came to take Karl and me into custody. When we got up to dress ourselves, we had to empty our pockets and our *Berliners,* or "knapsacks", were confiscated. At the police station we were told we were suspects for what we, ourselves, would know best. We told the police commandant we didn't know what we were being charged with. "The better for you, then tomorrow you fellows may be free once again."

Karl and I were given pallets, 2 blankets and a cell. The door slammed shut and there we were, stuck in jail. The following morning at 8 a.m. we were given a mug of coffee and a piece of bread. At 11 a.m. we had a case hearing. We were then returned to our cell. When we again asked why we were in custody, we were given no answer. At noon we received a bowl soup and some bread. At 7 p.m. we were again given some bread and coffee. The following morning at around 10 a.m., we were taken to a big courtyard. There was a water pump and a basin where we were able to wash ourselves. Two prison guards stood nearby. When we finished, I stepped forward and told the guards we were not going to move until we'd been told what we'd been accused of. The policemen looked at one another then told us we were Meissen gambling suspects who had duped a linen merchant out of 45 daalders/thalers. Due to this suspicion, the police were sending us back to Meissen. On our way to the station, Karl started to cry. When I asked him why he was crying, he said, "When my mother would hear about this, that we were transported to the prison in Meissen, she would not survive it." After giving this a thought, I too started to feel queasy.

By the time Karl and I reached the railroad station, the police escorts began to have pity on us. They started to believe that we were innocent and offered both of us a large glass of beer. The train arrived. We got on the train and after some light humor and talk we were back in Meissen. This time we would have a chance to see the beautiful citadel, The Albrechtsburg, from the inside. After we walked up the 288 steps to our cellblocks, we were each placed in our consigned cells, numbered 14 and 17. Both cells had a beautiful view of the Elbe valley. But just like in Döbeln, we were given the typical jail/prison food. The following day was Sunday. We were glad that it was raining all day. I asked for a book to read and the jailers gave me a good one. The day went by quickly. During the night it was pretty noisy. One could hear the 4 rows of rattling chains along the bottom of the Elbe that were pulling the ferryboats.

The next morning I asked the jailers for a job. They gave me a barrel of lentils to pick provided I did a good job. I would earn 4 pfennigs. It would take at least 8 days to complete the task. At 10 a.m., Karl and I had our court hearing. The judge was skeptical of what we had to say. At 4 p.m., we were again summoned before the judge. But this time in the courtroom there was also a well-groomed short gentleman. The judge addressed the fellow, "Mr. Marks, do you know these two people?" "No." "Have you ever seen these people?" "No, sir." "Aren't these the two people that were gambling at your place?" All of a sudden, the short little fellow became a foot taller and averred to the police, "How could you arrest such two nice fellows?" He thought he had done a good job of describing and depicting to the police the exact profile of the culprits. "A shame! A shame!" Mr. Marks cried out. The judge then had Mr. Marks sign a statement and dismissed him. Mr. Marks then left the courtroom while continuously uttering the words, "Shame! Shame!" We never saw him again. Then we, the accused, were given a statement of our innocence and were released from detainment. But I complained to the judge that we were unjustly confined for 4 days and insisted on some compensation. The judge started to laugh and said we'd be given 17 groschen for travel money.

Karl and I ran like two hares down the stairs and off the mountain straight to number 329 Rosengasse. Upon our arrival we were received with open arms because Elise and Emily believed in our innocence. We enthusiastically celebrated late into the night before finally going to bed. The following morning, Elise showed me the *Meissener Newspaper* and pointed to the large insert: "Again gamblers have surfaced, this time not from Berlin, but now they were meat packers; Hubert Dielen from Straelen, Kreis Geldern from Pruisen and Karl Stroh from Florsheim near Mainz. These three men swindled a linen merchant out of 45 thalers." I grabbed a pair of scissors and clipped out the article as a souvenir. With this news item floating around, it seemed better to get out of town. We were also getting too friendly with the two ladies. I was getting the impression that Karl was falling in love with Emily and was going to announce an engagement. So we went on our way again, back to Döbeln.

At the inn in Döbeln, the innkeepers were quite inquisitive as to our late encounter with the police. They ended up offering us a free night stay. After all, we had hardly gotten our money's worth from the previous stay when we were unexpectedly awakened by that false arrest. The following day we took the road to Leipzig. In Leipzig, we stayed at *Das Alte Gildenhaus* inn located on Hospital Street. Projecting from the inn was an antique decorated shingle that depicted one butcher holding an ox while the other butcher heaved the hammer that subdued the ox. Written underneath was a statement in German: "The ox will fall if hit correctly on the first blow, giving us the rest of the day off."

The restaurant was crowded and the hotel rooms were well occupied. We sat down and ordered coffee. We had our own bread and sausage. The innkeeper asked for our passes for the police surveillance, which we of course handed to him without hesitation. It wasn't long before we were telling of our recent encounter with the police. The story wasn't finished when two men entered the restaurant and asked for Hubert Dielen and Karl Stroh. I said, "Watch out, here we go again." One of the men came towards me and I showed him our document from the Albrechtsburg, which proved our innocence. The man looked at the paper and said, "One can buy such trash for 2 groschen." He grabbed me by the collar and wanted to take me along to the local police station. I had a knife in the seam of my breeches just like one did in Bavaria and Tirol. I pulled out the knife and banged it on the table, "Butcher's blood is no buttermilk," I shouted in defense. The two detectives then left without explanation.

Just before this occurrence, Karl and I were having a conversation with a gin merchant named Friedrich Bummel, who was discussing whether we wanted to go with him to England as cattle escorts. There was ample work and the pay was good. Frederick Bummel was also annoyed by the approach of the two individuals toward Karl and me. So I asked Frederick, "Who were those two men without uniforms?" Then the innkeeper interjected, "They were the police. If it's true that you're innocent, I urge you to present yourselves at the police station. That'd be better for everyone involved." And so we went to the closest police station and an old gray-haired police officer met us with the remark, "You regret your transgressions so soon?" He then sent some telegraph messages and took us into custody.

At 11 a.m., the following morning, Karl and I were presented to the police superintendent who asked us a series of questions and then added that my reaction to the detectives was an obstruction of justice. I replied I didn't know those two men and they didn't show any identification. Also, Karl and I had already been falsely locked up for 4 days and who's going to make amends to us for someone else's error? The superintendent called over a police officer and said, "Book them until tomorrow morning. By then they'll have mended their ways." And so Karl and I again ended up in jail. The next day at 3 p.m. Karl and I were summoned before the judge, who was a nice gentleman, and two scribes. One scribe read the police report from which I was able to gather they had also telegraphed Straelen. The judge realized we were innocent of the gambling charges, but I still had the police obstruction charge. I again said I didn't know those men were detectives and they didn't identify themselves. But when the judge said I had pulled a knife and that meant a 3 month jail term, I felt myself getting pretty warm and uncomfortable. Both Karl and I started to cry. The judge got a little nervous and asked us to stop crying. Finally he decided to declare us free to go. As Karl and I were dismissed, we were given 15 groschen for travel money.

Back at *Das Alte Gildenhaus* inn, Karl and I told everything that occurred. Since we were the only guests left, they allowed us to consume whatever we wanted for free. After having a very enjoyable evening, we didn't have to pay for that night's lodging either. The following morning we discussed with the innkeeper the next phase of our journey. The innkeeper thought it'd be better to cross the border as soon as possible since we were registered on the police lists. But that idea didn't appeal to Karl and me. We preferred staying in Leipzig a couple more days to find a job or simply to replenish our sausage supply.

During this time I sent a letter home with the notorious newspaper clipping and a photograph of myself. I wrote that after Berlin I would journey to Hamburg and Bremen. I would then reach Straelen via Köln. From Köln, they would hear from me again. I hoped to find everyone safe and sound when I got home. I also wrote how the last couple of weeks had been rather disappointing but that is part of the adventure of being a wandering journeyman. Something one enjoys talking about at a later date. I also sent a letter to the Berlin general post office, where I expected to receive my parents' response a couple of days later.

Karl and I finally left Leipzig to journey to Torgau, Wittenburg, Trevenbritzen and so on until we reached Berlin. We approached Berlin via the dilapidated section of town and the closer we came to the heart of the city, the worse it got. We arrived in the city at night and wanted to find an inn. We decided on *Zur Heimat* (tr., At Home). This was one of many Protestant institutions organized by ministers. We prayed together in the morning and in the evening. I was no longer a member with the Kopling house because I hadn't paid my dues with them for quite some time.

My first objective was to seek out my parents' reply at the Berlin general post office. My mother begged and pleaded for me to immediately return home. She didn't want me to return by foot but use the enclosed 15 German marks to come home right away. But that wasn't what I had in mind. I had yet to complete the route I originally planned. Not to mention that traveling in those days was all that easy. After all, a lot of traveling was still done by foot. Day after day I would go from door to door asking for bread until I found a job. These jobless spells happened many a time, usually lasting 7–8 weeks. One jobless spell even lasted 14 weeks. By that time your clothes get dirty and worn down. Then it is even harder to find work.

Karl and I then visited a couple of butcher shops to collect more sausages, before buying a large loaf of raisin bread. Close to *Zur Heimat* in Oranien Street we bought a *Berliner Weisse* (an ale) to top it off. We then had ourselves a good meal. To mention everything we saw would be too lengthy, besides you can get this information in a tourist brochure.

After being in Berlin and Potsdam for 12 days, Karl and I still hadn't found a job and our money was about spent. Our clothes started to look kind of raggedy so we decided to go to the stockyard. A livestock dealer told us one could always find high pay work in Hamburg. At that time there was a law that there had to be several attendants accompanying a train full of livestock. He looked like he was willing to take us along free of charge. Karl and I didn't need much time to think it over and promptly went along. Karl and I were first given a good solid meal at the train station cafeteria, which was in close proximity to the train. The train sounded as if it was ready to go. Before departure, Karl and I each received 10 rolls with sausage and a jug of caraway brandy from the cafeteria. That was a good thing because it was going to be a very cold night. When Karl and I found we had to stay with the livestock in the car, we picked a covered car that also had fewer pigs in it. Two pallets were thrown behind a trelliswork (wooden) fence. Our knapsacks with our provisions were on top of the pallets and a lantern hung on a hook. We then had to wait for the inspectors. Since we had to simulate 6 attendants, we rushed back and forth from one car to another to fool the inspectors who walked by the cars. Nobody caught on and so the train left on time. With a couple of old blankets from the livestock dealer, we made ourselves as comfortable as possible and tried to catch some sleep as we rode into the dark of night.

The comfort wouldn't last long. The pigs had begun sniffing at the provisions in our satchels and had started to dig in. By beating on the pigs as hard as we could, Karl and I were able to save our food. In the middle of the night the train stopped at a station and the stationmaster walked alongside the train to check on the livestock attendants. Karl and I were a bit curious as to where our livestock dealer was hanging out. There was one car coupled to the train, which was half passenger and half freight. Sure enough, there he was lying down on a bench all stretched out. He was wrapped in a fur coat underneath a blanket and snoring like an ox. Two blankets were still lying on the ground. Quickly, we snatched the 2 blankets that were lying on the ground and went back to our other blankets and knapsacks. We were now able to make ourselves comfortable in another compartment. First a good gulp from the jug and then we dimmed the lantern and went to sleep.

It was already light outside when Karl and I heard our boss coughing. Karl and I went to visit him in his compartment. Our boss was a jolly good guy of whom we were able to kid around with. Karl and I then took out our breakfast from our satchels but our breakfast had a pig smell. We were lucky that the boss had plenty of food with him. After we partook in our boss's breakfast, we told him about our previous experiences. He enjoyed listening to our stories. The time passed quickly. Soon the train stopped by a large railroad station and we again had to simulate 6 livestock attendants. The livestock dealer was quite happy that no one caught on to our scheme. Otherwise, he would've had to find and pay for 4 more attendants. The stopover was going to be about an hour, so the livestock dealer summoned us to the restaurant at the station for 2 large bowls of pea soup with a ham-shank and sausage. Karl and I had not eaten like this in a long time.

After Karl and I returned to the livestock car, we waited for the train to depart again. After which we roved about from car to car until we ended up in the compartment of the livestock dealer. While all of us were enjoying a fresh pipe of tobacco, Karl and I resumed our story telling with our boss. Where in the world Karl and I were, we didn't know. There was another long stop during the night where we were again each treated to a bowl of pea soup with all of the trimmings. Karl and I were getting accustomed to the world of livestock transporting. At daybreak, after a cold night, the city of Hamburg appeared on the horizon.

It must have been 8 a.m. when we arrived at the large livestock market. Around noontime, all the livestock were lined up in stalls. At the cafeteria, Karl and I could eat and drink whatever we wanted. The livestock dealer paid us 4 good thalers, shook our hands and wished us a good journey. At the large slaughterhouse, Karl and I asked everywhere for work but everyone referred us to the export slaughterhouse. But we decided to go first to *Zur Heimat* for lodging. There, we filled our stomachs with a couple pieces of sausage that some butchers had earlier given us and then went early to bed. The following morning, Karl and I saw the city as we looked for work. But all the jobs in Hamburg were already taken, even in Harburg and Genkstad. We were given sausages at prospective job sites, which kept us alive, but no work. So we finally went to the aforementioned export slaughterhouse. What a brutal workplace and bunch of massacrers! I asked for my pay that first evening and didn't return. But Karl thought he'd give it another try.

Not me, I went to the port. There was anchored a large passenger ship of the Hamburg-America line. They had an opening for a second cook. I signed on and could begin in 4 days. Back at the hostel, Karl was already waiting for me. He too had run off, but without pay. He regretted not going with me. We dashed off to the passenger ship office to sign him on too, but the office had already closed. In the morning, Karl and I stopped by a beer hall on the wharf. There we chatted with the loquacious, fifty-something year old, taverner of the beer hall. I told the taverner that in 4 days I would be leaving for America and that my friend, Karl, wanted to go too. The fellow then began portraying how old and unseaworthy the ocean liners actually were! He described how the ships would sink in the slightest storm and how horrendous the workers were treated. Karl quickly lost interest in working on a ship, but I persisted in inquiring further. When the next morning Karl and I heard the same stories from other people on the wharf, I too decided it wasn't for me. Karl and I went on to Bremen instead.

Karl and I weren't able to find any work in Bremen either, although we were able to find a job transporting livestock to Magdeburg and Halle. Halle was having its summer festival season and that made it difficult for us to find lodging. In Halle, Karl and I again went from store to store looking for work without success.

At the inn one evening, a very well dressed man approached Karl and me. He offered to pay each of us 15 marks per week to work at his job site and we could start the following morning. Now that was more than welcome. Our clothes had worn down to threads. Our shoes were in even worse shape and we were completely out of money. As promised, the fellow returned the next morning to pick us up. He appeared however a lot different, more common. We all first had a nip of gin before heading out to the city; on the marketplace in the center of the city stood a large tent, which we entered. The air inside the tent smelt awful. We were at a traveling zoo. When Karl and I asked what the job description was, the employer replied, "Whatever needs to be done." Karl and I would be able to decide amongst ourselves who'd do what. To start with, we would slaughter a horse; then we'd feed the animals. But the most important duty would be to keep all of the cages clean. That would all be done in this, the largest and most famous traveling zoo in all Europe. If we worked hard then we too would be able to travel to all the great cities of Europe, learn a lot and also become an animal trainer. I told the employer that would suit me just splendidly. To travel all around Europe would be simply fantastic!

Meanwhile, an enormous fat lady, with large gold earrings and a heavy gold chain around her neck, joined us. She too saw a great future for Karl and me in this traveling zoo. Karl and I told her we would immediately return as soon as we picked up our suitcases. But as soon as we got a breath of fresh air, we walked back to the inn and quickly left Halle (by foot) with not a penny in our pockets.

The road now led to Thüringen; Weimar, Erfurt, Gotha, and then Hanau. In Hanau, it was *fair* time again (to be exact, referred to as *kermis* in Dutch; the word is also in English dictionaries). At the journeyman's hostel, *Zum Freischütz* (tr., *To the Free Shelter),* it was so overcrowded that there was no room available. The innkeeper offered to allow Karl and me to sleep free of charge in the horse stable. He also gave us some free potatoes, two large herrings and a large pitcher of beer. Thus, having filled our stomachs, were we able to go to the fair.

At 10 p.m., the house servant escorted Karl and me to the horse stable. There wasn't much light but enough to notice that there were a whole lot of other people in our situation who were lodging there. One person was in a heavy overcoat, another wrapped in a blanket, and so on. Next to a bundle of old rags we found a place. The stable was adjacent to the dance tent, so falling asleep wasn't easy but we managed. Several times I felt as if a stick was poking my legs, but it didn't really awaken me. In the morning, after many of the other sleepers had already left, I got another jab from the stick. But this time, I grabbed the stick and yanked it.

Suddenly someone opened the door of the stable and as a result of the incoming light I was now able to see that I was holding the peg leg of a shrieking old woman. She resembled a witch. Now I understood what the bundle of rags was all about. The woman had covered her face with her raggedy clothes and had jabbed me by mistake with her peg leg. My yanking of the stick/peg must have been very painful to her. Then Karl and I recognized her partner. The couple owned the hand organ that was sitting in the corner. Yesterday Karl and I had helped the couple during their busy time with the collection of their tips.

We quickly reacquainted and the old woman was very pleased with her lodge companions. Her husband helped her stand up so she could hobble toward her hand organ as she held a bottle of brandy in her arm. She then grabbed a large snifter that was sitting on the hand organ and poured herself an ample portion of brandy. The second drink she offered to me, "I want you to know that it is made from real plums." I declined and suggested she serve her husband first, otherwise he may become jealous. The woman replied, "Imagine that! The instrument, the playing permit and the musicianship are all my doing, and if I could get a younger guy, I'd tell him (her husband) to go to hell."

After Karl and I had cleaned ourselves at the water pump outside, we went to the journeyman hall in high spirits. We each paid a kreuzer for a cup of coffee, another for sugar, and one for bread, which we then combined with the sausage we still had. We considered it a good breakfast. Karl and I spent the whole morning looking for work to no avail. Back at the *Zum Freischütz* hostel, the innkeeper asked whether we were going to stay longer. I said, "Sure, but not in the horse stable again." My partner, Karl, was not following the story and almost ruined it for us. I had the impression that since the innkeeper was obligated to lodge us journeyman for practically no fee, he wanted to get rid of us. The innkeeper persuaded Karl and me to accept a good noontime dinner and 30 kreuzers each, provided we would leave.

At 1 p.m., Karl and I were on the road to Frankfurt. In Frankfurt we almost didn't find shelter until we came upon a hostel named *Zum Heiligen Geist* (tr., *To the Holy Ghost*). I didn't like it there at all. When I went downstairs in the hopes of finding better lodging somewhere else, I saw something I'll never forget. In the main barroom were about 25 young women looking for an escort. At that time in Frankfurt, women weren't allowed to be out on the street alone after 9 p.m. Karl had already picked himself a girl, but I was suspicious and told the innkeeper that I didn't feel well and was going to bed. Everyone laughed at me but I nonetheless went to my designated bed.

The so-called bed was actually just a box, half-filled with straw, a dirty blanket and some rags as a head pillow. The beds were situated in a corridor, so sleeping there wasn't easy. There was noise all night long with singing and boozing. Then at 7 a.m., when it was calmer, I got up to go outside. I encountered some insults on my way to the door. It had always been my custom to go to Mass every morning when possible, so today was no different. At church I quickly fell asleep. The organ playing woke me up and after I went outside, I began looking for a *Kolping Journeymen Organization,* which I found soon enough. After I described my recent experiences to the priest in charge of the organization, he empathized with my predicament and rebooked me as a member of the organization. But if I hadn't found work after a few days, I would have to move on. Such were the rules of Father Kolping, the originator of the Journeymen homes.

After a few days, I went toward Wipperfürth via Bad Homburg. The next day, after walking for a couple of hours, I arrived in a small village and asked for bread at the first house. The snow laid thickly on the ground. But oh my, I had knocked on the door of a policeman. He gave me a choice, either I spend 3 days in jail or leave the village. I chose the latter. Close to the outskirts of the village there was one more house. I went straight to it after making sure no police were around. The homeowner was a baker. He had two friendly young girls who each made me a good sandwich with lots of filling. Just when I wanted to leave, whom do I see? That awful policeman again. I took off like a deer, but now in a snowstorm. After a while, I looked over my shoulder, but still saw in the distance the policeman following me. He then brandished his saber to which I countered by brandishing my stick. The policeman then grinned as he turned around and vanished into the blizzard.

In Wipperfürth I had to spend the night but first I paid a visit to the local butcher. The butcher took pity on me and gave me some money.

The following morning I was up and on the road at 8 a.m. because I wanted to reach Lennep before dark. I thought I'd find a fellow from Straelen there. By about 2:30 p.m., I had reached Lennep and inquired at the local inn about a Mr. Gooszens. Although I was told he no longer lived in Lennep, it was mentioned that another person from Straelen did live nearby. He was a tailor named Hein Mai. So I first went to Hein Mai at the tailor shop. Arriving at his room I recognized him right away, but he didn't remember me. I asked him if he didn't want to remember me, in which case I would leave. When I told him I was Hubert Dielen, Heinrich put his hands together and said, "What a sight you are!"

I told him that I hadn't been working for the last 4 weeks or so and that the cold and snow only compounded my current state of misery. Heinrich went into the next room to tell my story and then asked me to come in to repeat it. I was then given a good meal.

Heinrich and his patron had taken me to heart. First they cut my hair and shaved off my beard and then gave me a bundle of Sunday clothes. That was great. Now I needed some shoes. It wasn't easy but I did manage to borrow a pair. I then went with Hein to look for work. In the meantime, an apprentice repaired my old suit and had a cobbler repair my old shoes. Since there was no point in my continuing the journey in this kind of weather, I had to find work (in Lennep). And sure enough, Hein, who was well known in town, was able to get me a job with Mr. Ewald Hasselkuss.

Back at the inn, Hein and I had a good meal and before I was about to go to bed, I remember feeling very warm before becoming delirious. When I awoke, a doctor, a priest and Hein were standing by my bed. When it dawned on me what was occurring, I started to cry and asked for my mother. The doctor then looked at me and let out a good laugh. He said, "This young fellow has been subjected to too much cold and misery. He'll soon feel a lot better." Then the doctor and the priest left while Hein stayed behind a while to give me every now and then a sip from a drink that burned my throat. The following morning at 9 a.m., the doctor and Hein returned to see me. Hein informed the doctor what had happened during the night. I responded, "I feel better but I'm extremely hungry." The doctor declared me *well* but suggested I take it easy for a couple more days. As soon as the doctor had left the room, I jumped out of bed. Hein and I went downstairs where we were served coffee, bread and bacon.

I then went to my new boss. He was a bit apprehensive at first, but Hein convinced him I was all right and could begin. The wife of the boss first came with a warm, woolen sweater, long underwear, heavy warm socks and a pair of leather clogs. As soon as I was completely outfitted, I sized up my new work territory. But it was so filthy, I was about to walk away. Even I, an apprentice, instantly knew that my new boss didn't know much about his business. So we agreed I'd first get the sausage-making kitchen in order. My boss left it all up to me. First, I went to the store to buy a brush and plaster to begin plastering. The kettle of water had heated up enough that as soon as I had completed a second coating of plaster, I was able to begin scrubbing the dirt off of the benches and floor. It was quite late by the time I finished and everyone had already gone to bed.

The next morning, the boss and madam came to see how things were going and were quite delighted. Trautchen, a nice appealing maid, had given the shop a good cleaning. It looked spick-and-span, but now what? It seemed my boss barely did anything and simply purchased any needed merchandise from other butchers. Although Mr. Ewald Hasselkuss and his wife were well liked in Lennep and had plenty of customers, they didn't seem to make any profit. They both came from families of means, so they could survive. I came up with an idea. It was decided the boss would go the following morning to the livestock market in Elberfeld to buy two pigs and some beef. We would place an appealing advertisement in the paper and hope that would suffice. The wife was all for the plan and promised to assist where needed.

I began preparing the sausage kitchen and lining up the sausage casings. At 8 p.m., I had enough and went to the kitchen, where Mrs. Ewald Hasselkuss, Trautchen and Hein were busy talking about me.

After smoking a pipe, I went off to bed. Hein had to work overtime all through the night. It was close to Christmas and the time he had spent with me he had to make up. At noon, Thursday, the boss returned from Elberfeld and I made sure the water in the kettle was hot. At 2 p.m., the pigs arrived. At 4 p.m. they hung in the store neat and prepared. Mr. Ewald Hasselkuss and I then went along all the nearby butcheries to buy more racks of beef. It soon became apparent to me why they were so happy to see Ewald. He paid too much. I was able to bargain down the price a little but not much.

When the meat arrived at the store, Mrs. Ewald Hasselkuss had second thoughts. It was quite a lot of money. But when I explained I would need much more meat by Saturday and would've been more than happy to loan the money if I had it, Mr. and Mrs. both laughed heartily and gave the money. We created an attractive advertisement for the coming Saturday before we all said "good-night." By 4 a.m. the following morning, I got the fire started in the sausage kitchen.

A pig was now cut up, so I was able to make ham, sausage, salami, gotha, and then wieners and frankfurters. At 7 a.m., Trautchen had brought me a hot cup of coffee. When the boss appeared at 8 a.m., I had already done quite a lot of work. The boss helped me fill the remaining casings and then we put the sausages in the smoker. The boss was supposed to select the boiled meats for various kinds of sausages but that didn't work out too well. So I decided to let the boss do only the minor things such as buying the salted tongues, fat, spices, and so on. Mr. Ewald Hasselkuss took his time doing his assignments. He probably had a drink or two at the various establishments on the way. He preferred that to working in the sausage kitchen.

The sausage kitchen with all its machinery was now up and running and I was enjoying my work. Trautchen was right there whenever I needed some assistance. Around 4 p.m., I had enough sausages completed. Now it was the madam and Trautchen's turn to organize and set up the store. Large oblong and circular plates were filled with various kinds of sausages. While we were doing this, the newspaper arrived with the announcement of the new event. Then Holland was in distress (an idiom which can be rewritten here as: This put Mr. Ewald Hasselkuss on the spot.). Suddenly he had something else to do but the madam asked him wasn't it time he put on a white apron for once. I had a good laugh.

Soon some people stopped by to do some window-shopping. Just when I was placing a few of the trays with sausages in the shop window, some of Mr. Hasselkuss' friends entered the store. First they admired the neat appearances of the madam and

Trautchen. The two women did look pretty sharp. Then Mr. Hasselkuss' friends admired the merchandise. Mr. Hasselkuss was speechless. More of his friends stopped by and everyone was offered a slice of the excellent sausages. It didn't take long before the store was packed with people. Trautchen quickly fetched a large basket filled with soft rolls at the bakery while I heated some wieners and frankfurters in the kettle on the stove. As the German marks flowed, the Madam was barely able to maintain the financial transactions. The reopening was a great success. My good friend Hein also stopped in but it was so crowded I didn't notice him amongst all the people.

We closed shop at 10 p.m. Then Hein came to the family quarters to share a fine glass of wine. While we were toasting the great success of the day, the boss all of a sudden became very emotional and started to cry. The Mrs. thought it quite amusing. As for myself, I just wanted to go to bed. The following day was Saturday and German bratwurst sausages and stuffed meat rolls were on the menu. Those items lasted only until the afternoon and then Mr. Ewald Hasselkuss had to go out and buy more supplies. We were relieved when that busy day was finally over.

On Sunday, we first went to church but returned to the store afterwards to clean up the sausage kitchen and the rest of the store. Then I wrote a long letter to my parents and also a letter to Bayreuth where I still had my suitcase with clothes. Since I didn't have a decent outfit, I didn't go out. I instead spent the rest of the day inside playing cards with Heinrich.

Monday, Mr. Ewald Hasselkuss again went to Elberfeld and came back with 5 pigs and a young steer as he promised. Otto, the boss's brother, helped us with the butchering of the animals and before dark, all the meat neatly hung side by side in the store. The business continued to improve and a huge strong guy was hired to be my assistant. My assistant and I were able to process on average about 10 pigs and one steer every 2 weeks.

When my suitcase with my better suits arrived a week later, I went out at daytime. The village soon gave me the nickname *the spiffy one.* One can understand how I saw this time period in Lennep as the happiest time of my life. My parents wrote me a letter in which they thought it'd be nice if I would start making plans to return home. Mr. Hasselkuss wrote my parents a letter appealing to let me stay a little longer and they wrote back giving their approval.

The winter season was over. I soon became known throughout the area since I went everywhere. I was good at dancing, singing, theatrics, and especially storytelling that pertained to my adventures. As a result, I became quite popular. My mother early on imprinted in my mind the hazards young women could present to young men. But Mr. Hasselkuss, my boss, endeavored to get me on the wrong track. Across from our store lived an ugly looking rich widow, Frau Böckenbach, with her two daughters. The older daughter was a 23-year-old redhead. The younger daughter was a 20-year-old blonde named Emma. Beside our house was a path that led directly toward the beautiful garden of the widow. Soon I got the feeling my boss and his wife had been talking about me to Emma. And so it wasn't long before Emma, passing through the garden, stopped by my window to chat with me. Emma wasn't exactly the most beautiful girl, but she was agreeable and friendly. Naturally, I'd sometimes walk along with her to the garden, which was full of beautiful roses and other flowers. Of course, one day the old lady met me there. She wasn't too thrilled to see me, but I explained I had quite an interest in flowers and my parents also had a large garden, which I dearly missed, at home. We had about 200 rosebushes, I boasted. This seemed to mellow her somewhat.

On a Sunday, the boss invited me to go with him to a large summer garden, the Knoesthöhe (tr., gnarled hill). We left around 5 p.m., and reached the park about half an hour later. Concerts and balls were often held there. After we arrived at the garden, we looked for a good spot and ordered wine. We were barely seated when the Böckenbach family arrived. The older daughter was with her fiancé, an engineer. He was even more homely than the two daughters. They seated themselves right next to Mr. Hasselkuss and me. As Mr. Hasselkuss and I were introduced to the engineer, our tables were shoved together. The engineer and I conversed seriously at first until he became very interested in my adventure stories. I was neatly dressed and made a good impression, which softened even the widow.

The dancing had started and the engineer asked whether I danced. I said, "I sure do, but I don't know any ladies (that dance)." "Well, there are two ladies sitting right across from you who are good dancers." So I bowed before the old lady and asked, "May I dance with Emma?" She then nodded and Emma and I happily took off toward the dance floor. Mr. Hasselkuss danced with Emma's sister. When we returned from the dance floor, we were complimented on our dancing. When everything was over, we all walked home together. Once we had reached our destination and said our good-byes, the engineer asked me to stop by tomorrow evening for a cup of tea. Since I still had a lot of work to do, I hesitated on agreeing. But then Mr. Hasselkuss said, "If Hubert wants to go, that'll be fine with me." So I accepted the invitation. Later on, Mr. and Mrs. Hasselkuss delighted in remarking, "How such an invitation was something I wouldn't find in Straelen."

I pressed the doorbell the following evening at 8:30 p.m. A woman servant opened the door and showed me to the living room. The servant took my coat and hat and offered me a seat next to Emma and her mother. We at first talked about all the things that happened yesterday and critiqued the music. The tea was quickly consumed and we switched to a bottle of wine. I knew wine consumption was dangerous territory and planned beforehand what to do. The time ticked away and it was 11 p.m. before we knew it. When I saw another bottle of wine uncorked, I said that it was time for me to leave. The engineer, who already had a few drinks too many didn't think my leaving was much fun. But I said, "I may be young, but I know what I have to do. Otherwise, I would've been lost amongst all the losers long ago." The clock on the mantelshelf chimed 11 p.m. I emptied my glass and said good-bye to the engineer who was leaving in the morning. I thought about my mother's admonitions as I went home with a heavy head.

The following morning at 7 a.m., the engineer stopped by the store to say good-bye again. As he shook my hand, he said he didn't believe I was as upstanding as I had presented myself last night, but nonetheless, it made a good impression on the Böckenbach family. At breakfast, Ewald and his wife asked how everything went last night.

Meanwhile, Hein had gone back to his home in Straelen for 4 weeks and also visited his brother who had formerly been the pastor of Witten. When Hein returned, Mr. Hasselkuss informed him of all the late happenings. Hein seemed pretty jealous and I had the impression he didn't care too much for me anymore. That bothered me a lot.

One morning, the boss called me. There was a telegram for me from Straelen. The telegram read, "Mother very ill, send Hubert home immediately." According to the train schedule, I could catch a train in an hour and be in Nieukerk by 9 p.m. that same evening. Then it would still take an hour and a half walking to Straelen. So, Mr. Hasselkuss and Hein took me to the railroad station. The train ride to Nieukerk seemed to take forever.

When I finally arrived at Nieukerk with the little suitcase that Hein loaned me, I saw Mrs. Pasch waiting for me. I asked her whether my mother was still alive. "Sure," she said, "but she's very weak. Come along, I have a chaise waiting. Get in and I'll take you to Straelen." I asked her how she knew I'd be arriving. She said Hein sent a telegram. How could I have so misread Hein?

Mrs. Pasch cracked the whip and off we went. It took us 30 minutes to get to Straelen. We came across a group of women who had just returned from St. Ann's chapel after they'd been praying for my mother's recovery. Mother heard my footsteps and sat straight up. At first it appeared as if Mom didn't recognize me, but when I spoke she grabbed me with both hands and wouldn't let go. Father sat nearby and was crying. She said to him, "Don't cry anymore. Now Hubert has returned, I won't die." Dr. Brum, the house doctor, also came over to the house. He examined my mother quickly and decided she was improving. He advised her to take it real easy and left again. In the mean time, Mrs. Pasch said her good-byes and went back to Nieukerk. After talking for a while, Mother said, "Your coming home has been my best medicine." Downstairs in the living room, my brothers and sister wanted to know a little about my adventures before we all went to bed for the night.

After everyone had a good night sleep, we children went to church in the morning. At 9 a.m., Dr. Brum gave Mother another examination and said, "You seemed to have been worrying about your son." Mother replied if this illness continued, she wouldn't live much longer. Mother had something wrong with her stomach and would vomit as soon as she ate something. Drika, Louis and I asked Dr. Brum whether he would mind if we had Dr. Benz examine Mother. Dr. Benz was the son of a previous boss of my father.

He had a good reputation but didn't seem to be a friend of Dr. Brum. Dr. Brum said it would be fine with him if our father didn't mind. I assured him that I'd take care of that. "Will you be here at four o'clock, Dr. Brum?" "As usual," was the reply. So I immediately went to talk to Dr. Benz. Dr. Benz just happened to be stepping out of his house for a ride. He was very friendly and promised to also be at our home at 4 p.m. Father had yet to be informed of this, so I asked Father whether he objected to Dr. Benz also taking a look at Mother. Father said if Dr. Benz wanted to do that, it would be fine with him.

At 4 p.m., Dr. Brum arrived. Dr. Benz was already conferring with Father. Both doctors greeted one another and went to Mother's room. Mother was of course told about the second doctor. Everyone had to leave the bedroom because Dr. Benz wanted to examine Mother. After half an hour, both doctors went to an adjacent room to discuss the situation. Finally, Dr. Benz wrote a prescription and said he'd return later in the evening. Mother had to take once every hour, 5 drops of medication with a spoon of sugar water. Dr. Benz returned at 10 p.m., took Mother's pulse and asked whether she'd had to vomit anymore, then he left.

The following morning after Mother had had a good night sleep, she felt much better. She didn't vomit anymore and Dr. Benz, who'd already stopped by at 7 a.m., returned again at 9 a.m. with Dr. Brum. Both doctors agreed that Mother could slowly return to eating. Daily, Mother progressed. She was now able to hold her food down. By the second day, she was already able to get up and sit at the window for about an hour. I kept Mother company almost the entire day. I told her all about my adventures. Due to Mother's improving condition, I decided to write a letter to Mr. Hasselkuss in Lennep mentioning the possibility of my returning to work in about fourteen days. After a few days, Mother was able to walk downstairs. After a week, she was able to walk to church and to our garden.

The following Wednesday, I sent a telegram to Mr. Hasselkuss in Lennep: "Will arrive tomorrow." Although I packed my bags to return to Lennep, I had to promise my mom I wouldn't stay away so long anymore. Only then was I allowed to return to my beloved Lennep. Even though my assistant had done his best in maintaining things, it was high time for my return. I was received with open arms and began with new energy, but I wasn't as enthusiastic anymore. I kept thinking about home. I told the boss to start looking for another employee, but nothing happened. So I decided to find a replacement myself. I found a nice young fellow and helped him get oriented for a couple of days. Then on Friday I picked up my pass at the town hall. Sunday evening, I went dancing one last time with Emma and Hein at Knoesthöhe. Then on Monday morning, without any further good-byes, I went with my suitcase to catch the 7 a.m. train toward Straelen.

I journeyed through Elberfeld on my way to Düsseldorf. In Düsseldorf, I wanted to visit an inn I had once frequented. There, I unexpectedly came across an old buddy I had traveled with in Strassburg. Within an hour he had convinced me to make an excursion through Belgium. So I retrieved my suitcase from the station, put my better clothes in it and delivered the suitcase to a forwarding office. I wrote a letter home saying I'd return after first visiting Belgium. I was now traveling by train with my friend to Köln. In Köln, we found ourselves wondering around some and before I knew it, I was back on the road as a journeyman. The journey took us first to Kerpen, Düren, Aachen and then to Verviers, Belgium. But, oh my! Here came the police again. Before we knew what was happening, we were returned back to the other side of the German border because we didn't have a passport. Now I had had my fill of journeying about and took the road back toward Straelen. My travel companion, who wanted to continue on toward Antwerp, stayed behind.

As I went my way, I arrived at Mönchengladbach. There, Rudolf Schiffer, who often traveled to Straelen to buy pigs, offered me a job. He knew our family very well but didn't recognize me. I asked the forwarding company to mail my suitcase with my better clothes, so I could dress up a little and make a better impression. My new boss was often on the road and I got along fine with his mother and two sisters. Schiffer was a 35-year-old bachelor who had a real zest for life. He was a smart dresser and had a good appearance.

After a couple of weeks, my boss wanted to know where I really came from. I told him I was from Württemberg and had been traveling around for a longtime. I showed him a fake pass which older journeymen always kept within the lining of their coats. Schiffer was suspicious. A week later, when Schiffer went to

Straelen to buy 2 pigs from my brother Louis, Schiffer casually asked Louis where Hubert was hanging out. My father and Louis told Schiffer the latest news about me and that I was probably in the vicinity of Antwerp. Schiffer smiled at them but didn't say anything. Schiffer returned home that night with the 2 pigs he had bought and I began butchering them.

After I had finished my work, Schiffer invited me to go out and have a drink. I thought it was a little late but he insisted. Schiffer ordered a bottle of wine, filled the glasses and said, "Cheers to you Hubert Dielen, and not Robert Mertens. The 2 pigs you had just butchered, I had bought this morning from your brother Louis. I had also seen the rest of your family and they were of the mind you had gone to Antwerp. But your family really wished you'd come home." That did it. I couldn't keep my charade any longer and told Rudolf Schiffer the truth. After Rudolf and I finished the bottle, we returned home. There Rudolf's mother and two sisters were doing their needlework. Rudolf told them what he heard from the Dielen's in Straelen and they didn't know where their son Hubert was. The sisters heard from (Hubert's sister) Drika's last visit that Hubert traveled a lot but was then residing in Lennep near Elberfeld. I sat like a stone in my chair without saying a word. I felt everyone's eyes upon me. Then Rudolf immediately said, "This Robert Mertens here is actually Hubert Dielen." To top it off, I learned that Rudolf Schiffer was courting my sister Drika. My mask had been completely pulled off. So I told them all about my adventures. I talked late into the night as everyone listened.

Since Rudolf had to go again to Straelen the following Monday and was able to find a good journeyman to replace me, we agreed I should go along to Straelen and bring a closure to my adventurous journeyman lifestyle. So on Sunday, Rudolf and I went once again to town for a last hurrah. Monday morning my suitcase was on the wagon and I said good-bye wholeheartedly as I warmly embraced these wonderful people. By 11 a.m., we were close to Straelen. I jumped off the wagon while Rudolf drove on toward the house to ready my family for my arrival.

When I entered the house, everybody was standing around Rudolf listening to what he had to say. I don't know what kind of robber stories he was telling, but when the family saw me enter the room, my mom ran over to me with outstretched arms and laughingly gave me a warm hug. She would not let me go. For my homecoming, my family put on a great feast and Rudolf stayed to join in the celebration. Louis, who ordinarily didn't have much time for me, stood close by.

After the meal was over, Rudolf put the pigs he had bought from Louis on his wagon and departed for Mönchengladbach. At home, due to my return, the day was a holiday. My father obviously felt proud of me. First, he wanted to show me his garden, which was located next to Sant, a nearby village. By evening, we had gone from one pub to another. After having had a lot to drink, we went home to sleep. How great it was to be back home! "Home, sweet home."

IV. The German Cavalry

The entire family went to Mass the first thing the following morning. After Mass, Louis suggested I take the day off. I laughed off that idea. In a brotherly manner, we shared the work and I went right on doing my part. But only a week later, I had to register for the draft at a recruiting station in Geldern since I'd been traveling around as a journeyman when my last draft registration notice arrived.

Aloysius van der Velden, who worked at a tannery and who was a boarder at my parents' home, asked me to deliver a letter to his parents in Geldern. His parents ran a grocery and dry goods store on Issumer Street. So that was the first address I headed for. I was well received by his parents. I was even better received by their daughter, Alwienchen, the sweet pretty girl that ran the store. The old gentleman Mr. van der Velden offered to go with me to the recruiting station since he knew well all of the gentlemen there and thought he could help me avoid any sort of punishment (I may have incurred by my belated registration).

At the command station, I had to undress for a physical examination by the doctor. A major, who sat next to the doctor, then barked, "First Pommern Uhlanen, move on." Then Mr. van der Velden and I left.

At the van der Velden's home, the table was set for a full meal, which set the pace for a most enjoyable afternoon, including a long walk with Alwienchen. This friendship gave me good reason to invite Alwienchen to the fair in Straelen. Father and daughter brought me to the train station and said, "Till Sunday."

Back home, I told Mother how wonderful the reception at the van der Velden's had been and I couldn't avoid not returning the favor. That's why I had already invited Alwienchen to the fair. Mother laughed and said, "Yes Hubert, that's the way girls are." I also wrote the Böckenbach family a letter apologizing for not saying good-bye in person when I was leaving. I hoped to meet them again in the future when I'd repeat the apology. They were such a nice family.

To my great satisfaction, Louis and I were now able to work well together. Before the fair, we had a lot of work to do. Lots of sausages had to be prepared and everything went smoothly. Aloysius van der Velden went home on Saturdays and I asked him and his sister Alwienchen to come to the fair on Sunday. The autumn festival had quite a crowd and the Dielen's had a lot of guests as a result. Mother said, "It's a good thing we just extended and renovated the house, otherwise I wouldn't have known what to do with all the guests."

Since we hadn't slept very much in the last few days, Sunday was spent resting, updating our account books and tidying up. Monday at 10 a.m., the table was set for our guests from the fair and some coffee made. The Pasch family of Nieukerk was the first to arrive. They stopped by to say, "Hello." After that, I strolled with my two dogs, Caesar and Nero, to the station. There, I saw Aloysius and Alwienchen from Geldern get off of the train. Alwienchen looked so fashionable and city-girl-like that had I not traveled so much, I as a country boy from Straelen would have felt too awkward to be around her. As Aloysius, Alwienchen and I walked through the city; my two big dogs were jumping and barking ahead of us, at every door stood someone looking at us. Hein Mai had previously gossiped that Hubert Dielen had found a rich bride in Lennep. I wonder what they thought now?

When we got home, my mother was a little taken aback but my father thought it kind of neat. Alwienchen with her suitcase was ushered upstairs. Ten minutes later she returned and stood next to Mother at the stove dressed in a long apron. Mother was very pleased. Alwienchen had found Mother's soft spot. First, we all had a cup of coffee followed by drinks. Then we ate (our noontime dinner) and trooped over to Sant where a tent had been set up for dancing. The weather was beautiful. The girls were all nicely dressed. Alwienchen too had changed clothes but looked so chic, that my sister and the other young women seemed a bit jealous. I even heard them using the words, "Geldern bluff." But her pleasant disposition made her quite popular and that continued on into the dance tent where it was difficult to get a dance with her. She was all the time spoken for. That was fine with me, since it gave me the opportunity to dance with the other girls, such as Kate, Nelli and Drika Pasch.

By evening, we were home again. After supper, we dressed up for the ball at Lom. Alwienchen looked very beautiful and hung onto Mother's arm. Under a long coat, she wore a ball gown that had decorations similar to an uhlan. My first dance was with Alwienchen and I felt pretty good when I noticed everyone looking at us. Even Louis, who was usually quite reserved, didn't miss a dance. It was a wonderful evening for everyone. The festival eventually came to an end, Alwienchen went back home and everyone went back to work. Fourteen days later, Aloysius invited my sister and me to Geldern where my sister and I always loved to go and where we all had another great time.

Autumn came and one day I received my draft notice. I had to report for duty at 2 p.m. November 3rd, 1875 in Geldern. Until then, I continued working up to the last morning. Then I put my most necessary items into a small knapsack, gave a brief farewell and quickly left. Louis accompanied me to Geldern. At 3 p.m., the roll call was over and I was quartered at the van der Velden home. I then said good-bye to Louis from the van der Velden home. Louis returned home by train with a thousand best wishes for Mother from me. Since I had to report for military duty at 7 a.m., my host family and I went early to bed.

It was during my military service that my brother Johan became ill and returned home. He died in 1879. Something that Mother took very hard.

The following morning, Alwienchen and her father brought me to the assembling place near the train station. I was glad to get into the boxcar and leave as soon as possible. My companions were two farm boys who were also from Straelen and Piet Lanken who was from Geldern. The two farm boys, whom I really didn't know, were named Wänders from Brückske and Eickelpasch from the Land. Late in the afternoon, we arrived at the military barracks in Duisburg where we received a bowl of pea soup with bread. This was the first of the soldier's mess, which tasted pretty bad. At night, we were again transported like cattle in a boxcar. After traveling the whole night and most of the next day, we arrived at night in Diedenhofen (today it's Thionville, France). At the barracks, we were shown our quarters. Again we received a bowl of soup with bread and we were allowed to relax on our pallets. The night quickly passed because we were all dead tired from the journey in the boxcars. The following morning at roll call, the captain selected people for assignment. I was chosen for the second squadron of the first Pommern Regiment Uhlanen, which had been in Diedenhofen/Thionville since 1871. This regiment consisted almost entirely of Polish and others from Pommern. The Pommerns didn't get along with the French. Rhinelanders thus gradually replaced the Pommerns.

I was placed in room number 43, bunk number 7. Then us new recruits went to receive our uniforms, pack our civilian clothes and to receive a *night pot.* When we went to the canteen, this so-called *night pot* apparently was a food container with a coupon. At the canteen we were given 50 grams of meat and soup in the *mess-tin* as the *night pot* was also called. The food was definitely Polish. It was hard for us to stomach. After one hour, we were ordered back to the kitchen. Now we each received a 6 lb. loaf of bread. It was to supply us for 4 days.

On the second day, we were given our horses. I was given a brown horse named Panter. First we had to groom our horses, then groom and groom some more. Then we were allowed to take our respective horses out of the stables and walk them so the horse and rider could get accustomed to one another. In the evening, we all had to write our autobiography. When the captain read my autobiography, he started to laugh and wanted to see me. He thought my autobiography was quite interesting and wanted to get to know me a little better. After 2 weeks, we were allowed to visit the town under the supervision of the senior soldiers. By this time, we were able to march and salute. The food in camp can be described in one word, *awful.*

The meals and daily routine occurred as follows; every four days you would receive 6 lbs. of bread. That was all right. In the mornings, you had to rise at 5 a.m. and make your bed. After that, you'd march from the barracks to the stables in ten minutes. The horse dung had to be picked up by hand and the horse groomed until 7 a.m. The stable master then fed the horses. At 7 a.m. occurred roll call. Then at 7:30 a.m., we marched back to the barracks. We then went to the kitchen where we would receive the familiar mug of coffee and bread. Then we had to drill with our horses until midday. Every noon, using your coupon, you would receive from the kitchen a piece of meat and a large ladle of soup; one day pea soup, the next day brown bean soup, the following day lentil soup, and then black bean soup. On Sunday, we would have noodles with 4 oz. of bacon. Surely, the pigs in Straelen ate better. From 1–2 p.m., we returned to the stables for the horse vaulting exercises. Afterwards, we were given another cup of coffee from the kitchen and nothing else. At night, you could take another piece of your bread. It may be hard to believe, but that's the way it was. A person could manage if he received packages from home. If not, he was to be pitied.

For the moment it was still bearable. We had shabby clothes for daywear. Most of the uniforms were quite old. They had been worn during the last war. We had a hard time keeping them in shape, which we did at nighttime. The petroleum-fueled lights would go out much too early since there was so little petroleum. So we bought, using small change, a candle from the canteen. When a half of a candle remained at the end of the day, we would put the candle on top of the chest for the time being. The next morning, you'd see a *Polack* (Polish soldier) snatch the half candle and scrape it over his sandwich with a touch of salt and he had a delicious meal. It'd take too much time to bring up all the stories that happened everyday. But one thing was for sure, the Polacks all stole like crows. When a person was caught breaking open a chest, he would be laid over a chair and given a good whipping. But we Rhinelanders didn't report the Polacks because that would cause more problems. Then we Rhinelanders would have the whole room against us. The Polacks were the hardest to get along with.

After awhile, I wrote home about the bad food situation. In response, I was allowed at my father's expense to go to an inexpensive restaurant now and then for a meal. I also bought a new pair of riding breeches and was able to get a secondhand *Uhlanka* (jacket) from my sergeant. I turned the jacket inside out for a better look. So now I looked a bit more presentable. I also happened to be our squadron's right-flank man. I made an earnest effort to keep my affairs and belongings neat and tidy. My captain and the other officers had taken notice of this. In addition to that, my parents mailed to the sergeant a package of the finest sausages. Every month I received from home 15 marks and a package of sausages, butter or meat. Other folks also sent me some money by mail.

In spite of my relative good fortune, one day I became very homesick. So I went to the captain and explained my ill feeling. He happened to be in a good mood that day and after a discussion with the sergeant, we walked to the regiment's office. I was then given a pass for 7 days. I sincerely thanked my captain and got out of there as fast as I could. I was allowed to wear my Sunday uniform and could immediately depart for the train station. As a result, I didn't have to begin counting my furlough until I arrived home. When I arrived home, there was great excitement. Even the neighbors became curious. I showed my leave pass to a couple of disbelieving veterans. I quickly put the uniform in the closet and was ready to go to work; so passed the week, which flew by in no time. I then returned to Diedenhofen in a good mood.

First, I reported to the captain for duty and to offer my parents' thanks. At this time of the year in the barracks the worst was now over. The non-commissioned officers left us alone, the summer had arrived and that made our field service much more agreeable. I now had to learn how to swim. In 6 days, I knew all the important strokes and was allowed in the deepest part of the pool. After 14 days, I could swim so well that I was chosen to teach swimming.

One day at the Mosel (Fr., Moselle) bridge, I encountered two journeymen. Lo and behold, one of the two happened to be my brother Philip who was looking for work in Diedenhofen so he could keep me company now and then. The quartermaster, who witnessed our reunion, gave me the rest of the day off so I could be with Philip. My captain added another 2 days. So Philip and I decided to go to Metz (today it's in N.E. France) and Gravélatte. Soon thereafter, Philip found a temporary job. Then the big summer military exercises occurred and by the time that was over, Philip had moved on. The older soldiers were sent home and the young ones received fourteen days furlough.

When we returned, we saw a new group of recruits coming in, so that made us the senior soldiers. Now I received all kinds of choice assignments, such as fuel distributor, linen provider, and since I was a butcher, also kitchen controller. The kitchen assignment didn't last long because I was too critical. So I was transferred to be supervisor of the non-commissioner's table and of the classrooms. In the end, I was in charge of servicing the horse stables and the horse riding instruction. I didn't have any parade riding experience but was a quick learner and soon became pretty good at it. So, when we received an addition of chargers, I was the one to break them in. Soon thereafter, I was promoted to private 1st class.

Very proud private 1st class Dielen went home to Straelen for a 4–week furlough. I was able to enjoy this free time to my heart's content. I was now the owner of a formal uhlanen uniform and felt like a real gentleman. I, of course, visited my old haunts in Wankum, Herongen, Wachtendonk, Nieukerk and Geldern before heading back to my garrison. In Geldern, I unintentionally came across Alwienchen. I once again said how a continuous correspondence of love letters didn't fit in very well with a soldier's routine. The discontinuance of our correspondence was the reason she ended our relationship. Nevertheless, she insisted I stop by her house. Her mother eased up when I made it known I hadn't come especially for Alwienchen. Alwienchen was engaged to a rich widower named Frans Leeuw. Just the same, I reminded Alwienchen I was still too young to be tied down, since marriage was what she was really looking for. After all, I was just starting to enjoy the bachelor's life. I nonetheless wanted to remain good friends with the van der Velden family. I then stood, bowed, said my good-byes and left. That concluded that furlough. In good spirits, a well-filled suitcase, and a full wallet, I returned to my garrison.

One day, I had to ride the show-horse of lieutenant Levegow. Previously, an old sergeant had had a lot of trouble with this charger. Now it was my turn to subdue *Irene.* That was the name of the horse. Using a softer touch, I succeeded in a few days. Now nothing could go wrong for me because lieutenant Levegow was the captain's right-hand. Soon, I was training the horse of the captain as well. My life was really taking an upswing. All kinds of honor jobs were offered to me, such as the breaking-in of a horse in the beautiful surroundings of Diedenhofen. The two officers each paid me 10 marks per month for my efforts. I also participated in the great imperial cavalry exercises. Fourteen cavalry

Private First Class/(Lance Corporal) Dielen in Thionville (1877)

and artillery regiments took part in the maneuvers at Weisenburg (France: Wissembourg). The entire operation lasted 4 weeks. The weather was good, the exercises impressive, but the quarters were bad. That was to be expected. After all, the French were not too crazy about Germans. But I used my resourcefulness, which helped me in previous situations, to make the most of my circumstances.

In a small village near Weisenburg, I was assigned quarters in the home of a cantankerous cobbler for 14 days. His tall and thin wife showed me a very small stall with a bundle of straw for my horse and me. The horse felt fine, but I had to get used to it. I went to the cobbler's workshop to get acquainted with the man using what little French I knew. I asked him if there was anything I could do to help. His mustache trembled with anger when he replied, "There aren't any Prussian soldiers who can hammer a nail into a sole." I told him I was really a butcher but could still handle just about anything. This is where I played my card by saying I, just like him, was a compulsory German. I myself came from Holland. We then opened a map out on the table and I showed him exactly where Venlo was located. This explanation changed his mood somewhat. In poor German, he told me he had lost two brothers in the last war. Their house had been burned down and more and more stories were blurted out in a cursing manner, constantly condemning the Germans and their behavior. I felt relieved when the horn blew for the noontime meal. The cobbler said, "Let them keep their miserable food. I'll feed you," but that wasn't allowed. I told him that I also had to take care of my horse. At the canteen, we soldiers were allotted our three-day's worth of provisions.

When I returned to my ranting cobbler, the cobbler and his wife had an exquisite meal with a bottle of wine waiting for me. I thanked them and tried to excuse myself, but they insisted, saying I had to eat with them whenever I arrived at mealtime. After that, I also had to fill my pipe with tobacco from his pouch because in his opinion, German tobacco smelled awful. So, as long as I was in his house, I had to use his tobacco.

The afternoons were spent grooming the horse and taking care of the saddle, including the girths, leathers and irons. In the evening at 6 p.m. was roll call and inspection of man and horse. This lasted until 7 p.m. Then I'd return to my quarters where the table was again set for dinner. The cobbler had no children. He considered them too expensive and too much trouble. The cobbler asked whether I played billiards and whether I'd like to join him for a glass of wine at a local restaurant. "Sure, why not?" So we went to a hotel restaurant named *Bonnet d'Or* (The Golden Hat).

The first room we entered was the barroom, which had its bar counter and also a billiard table. Behind a few sliding doors was the dining room. I was a bit taken back, when I saw our captain with some other gentlemen in the dining room. As was the rule, I saluted and proceeded to the billiard table to practice a little and sip some wine. Suddenly, I was called to the table where the captain was seated. "The mayor, here, just told me you are quartered at the worst German hater in the whole area. Is that so? What do you know about this?" I didn't know what to say, so I kept my mouth shut. Then the mayor remarked how something must have occurred because the cobbler who had yet to say one word to a German and was just railing against the Germans to the innkeeper (at the bar) is now playing billiards with a German soldier. "Dielen, answer to this!" the captain shouted in a brusque manner. I told him the same story I told the cobbler and even mentioned the delicious entrées I was served. This was too much for the captain, who apparently hadn't eaten very well. He became very angry and wanted to put me in my place, but the snickering of the other gentlemen prevented him from doing so. Von Levegow, his first Lieutenant, remarked, "We've been receiving complaints from all sides about the quartering, but now I understand why Dielen isn't one of them." The captain wanted to know whether I had always been so shrewd. I told him how I was actually a butcher by trade and thus knew how to be resourceful. The captain didn't reply and while the other gentlemen were snickering, I returned to the billiard table. The cobbler played very well, much better than I. So the games turned over quicker than usual. Now and then the officers would watch us play but that didn't stop the cobbler with his epithets.

We soldiers were obligated to be in at 11 p.m., so I made sure to be on my way in time. Then the captain yelled out, "Dielen, you can stay till twelve, midnight!" But it seemed more sensible to leave because my billiard-playing cobbler was getting pretty drunk and started to curse the Germans louder and louder. In spite of everything, I must say I had pretty good quarters for those fourteen days, a lot better than what my colleagues had anyways. And when my day of departure came, I was given a well-filled knapsack.

That was just one episode out of many in my life as a soldier. The maneuvers were quite rough because the old emperor (William I) demanded that the riding be only in trot and gallop. As a result, there was a time when 35 soldiers ended up being hospitalized! But even army exercises come to an end and soon we were back in our barracks. With the exception of their non-commissioned officers, the Polish and the Pomeranian reserves went home.

One day I organized a pool amongst the reserves to provide a gift for the officers. Every reservist contributed every payday 25 pfennigs. We then ordered from a craftsman the creation of a pipe that was especially designed for our purpose. We also purchased a quirt in addition to some statues that were in the likeness of a reservist. These gifts were presented to the officers as souvenirs. In return, we received an extra night off with all the beer we wanted. I really started to enjoy my military life and after another 4 weeks of leave, I became a team leader. Since I mentioned several times to my superiors that I intended to stay in the army, they also made me recruit instructor.

The winter, which had by now begun, was a bitter one. I was now regularly training the chargers of the captain and lieutenant Levegow's in addition to my own horse. But I had all the freedom I could wish for and could come and go as I pleased. I even had a permanent night pass. So the winter went by without any unusual incidents except one; I had gone to the theater in town one evening but realized on my way back I had left my night pass in the barracks. I returned a little after 11 p.m. at the city gate and lo and behold, the night sentry, who was an infantryman recruit, wouldn't let me pass. I gave him a good kick and ran beyond him. The barracks I was quartered in was about 10 minutes down the road outside the city gate. The sentry at our camp heard me coming. He quickly opened the gate and once I was through, swiftly closed it. The three infantrymen who ran after me asked the sentry who he let in. "No one. Although there was a soldier walking by here, but I didn't let him in." The sergeant didn't believe him. It was a clear night and the sergeant had seen something. So the other infantryman wanted to speak to the commander of the guard. I was standing nearby hiding behind the sentry box. The commander of the guard, Jürgen, was from Krefeld and happened to be a friend of mine. He came outside and told the infantry sergeant, "No one has entered this gate. I have the key in my pocket." But the sergeant held steadfast, he wanted the gate opened. "If you all don't immediately get out of here, you'll all get smacked!" responded the commander of the guard.

At 9 a.m., I was alerted by my captain there was going to be a special regimental roll call at 11 a.m. He already suspected I was the culprit, so he advised me to shave my moustache and drink a couple glasses of vinegar from the canteen. I submitted to this prescription but it wasn't pleasant. I started to vomit like nobody's business and became as white as a sheet. At 11 a.m. was the roll call. The sergeant situated me in the middle of the third row although I belonged at the right of the first row.

Here as well, the infantry and cavalry didn't get along, even amongst the officers. The infantry sergeant said he could identify the uhlan by face but didn't know his name. That was why the commandant of the garrison ordered a roll call. The whole regiment stood at attention; everyone was tense. The three infantrymen of the city gate, together with the garrison commandant and a few officers arrived at the parade ground.

At first Jürgen and his sentries were questioned. But they kept a straight face. Then the first row of the first squadron was ordered to advance 3 steps forward. The three infantrymen, followed by the officers, looked over each soldier with no results. The first squadron could fall out. Then the first rank of the second squadron was ordered 3 steps forward. Suddenly, I started to feel quite uneasy. But then no one in the first rank was singled out either. The second rank was then ordered 3 steps forward. In the third rank, they focused on me for a moment, but then they moved on. And yes, sirree! The seventh man, soldier first class Bölton from Münster, was pointed out. The sergeant had seen me all right, because with my moustache and no vinegar, we could've passed as brothers.

Our captain, a clever guy, immediately intervened and asked, "Soldier Bölten, where were you last night?" "In the barracks." "What were you doing?" "First I put my uniform away and then I played cards until about 11 p.m." The captain then thundered, "Room 37, fall in!" "Who played cards last night with soldier Bölten?" Four men came forward. "Who witnessed that Bölten went to bed at 11 p.m.? All four raised their hand. The captain then turned to the commandant of the garrison and gave a salute. The commandant then furiously looked at the infantry sergeant, who then came straight toward me and said, "Then it's him!" Now lieutenant Levegow interfered, "That's not right. The sergeant points out someone and when it isn't him, he points out someone else. That can't be!" Other officers now intervened and agreed with Levegow. The captain quickly dismissed the second squadron. The sergeant ended up getting 5 days of severe detention and his two men 3 days. What a relief for me!

Back in my room, I was immediately summoned to the captain's office. There, all the officers of the squadron were present. "Where were you really, last night?" "I cleaned my clothes, played some cards and at 11 p.m. went to bed." The captain asked the sergeant, "Let room 34 line up." With a smile on his face, the captain asked the group whether my story was true. Only a concurrent muttering by everyone was the reply. Lieutenant Levegow said, "Captain, I'll bet you're proud to have a group with such solidarity!?" But the captain gravely bellowed at us the assurance that had the culprit been caught, he would've received a 3-year prison term. After this, I went straight to Bölten who was still shaken by the ordeal. He and I then went to the canteen and had quite a few drinks. The canteen manager said, "Now I know why the glasses of vinegar were needed."

Now that there were only Rhineländers and Westphälers in the barracks, the atmosphere had completely changed. Instead of having to constantly cleanup, we now spent the evenings singing and putting on skits. One day, I and some of the other fellows decided to ask the captain whether he would mind if we started a choir and if he'd also want to join. He declined since he couldn't sing very well, but offered to help. So I asked whether he could find for us a choir director who'd teach us 3 nights a week. The captain said he'd look into it right away and ask the regimental commander for assistance. That very evening, a sergeant musician presented himself to me. The following day, 40 of us soldiers stood around our new conductor. We had a serious rehearsal twice a week. But now, we needed books as well. I thought right away about our recently elected honorary president, the captain, and asked him for advice. He wanted to know whether I knew a bookstore. That wasn't our biggest problem, I said. The cost of the books was 1 mark each and our conductor needs 50 books! Conrad Storch in Berlin sold the books, books with mostly military and marching songs.

A week later, the postal service delivered a package to me. Voilà, there they were, 50 booklets as requested. The captain had paid for them. The attendance of the singers at the rehearsals was good. Those who hadn't participated for 3 weeks without a proper excuse weren't allowed to return. After a month, the other squadrons also started a music program. It wasn't long before the officers would come to the barracks in the evening, something that hadn't happened before, to listen to the musical programs.

In the spring, the garrison had a change of command. So we had to march through the city to the parade grounds. This time, though, there was a new order, "Singers forward!" First came the first and second tenors, and then came the first and second bass. Our conductor lifted his arm and then we began, *Wie lieblich sang die nachtigall,* translated as: "How sweet sang the nightingale," etc., etc. Even the French, who didn't know what was going on, came to their doors to listen. And so it was soon well known that the Prussians and the Pomeranians had left and were replaced by the lighthearted Rhineländers. In the coming weeks the relationship between the people and the military improved greatly.

My days now flew by because of all of the various activities. One day, the captain came by to talk to me and asked whether I was going to reenlistment for another year. When the captain would be promoted to major, I'd become a non-commissioned officer (N.C.O.). He had had enough of military life. When he then retired, I could be a steward on one of his two great estates; one manor had 64 horses, the other 50. He was a member of the Von Rudolph aristocracy and longed to return to the peaceful life in East Prussia. I told him I'd have to discuss the offer with my parents.

I promptly received a 4-week furlough and when that was almost over, I asked and received another 3 weeks. Seven weeks, that was unheard of! During the last 3 weeks, my sister Hendrika got married to an organist and music teacher, August Hendricks. That happened on June 11, 1877. But when I returned, (and my non-commissioned officer's uniform had already been made!) I had to tell the captain I couldn't reenlist. My father was getting older and my brother ailing. My father had already waited a long time for my return. Thus, my good life in the barracks quickly came to an end because my privileges were revoked and the N.C.O.'s soon heard of my not wanting to reenlist.

So now my duties reverted back to that of a regular soldier and I made sure not to do anything out of the ordinary. What surprised me though was that brunette Marie, the captain's domestic, who usually offered me coffee, wine, or cigars after my riding, was now even coming to the stables to offer me the treats and in a more assertive manner. I didn't understand it, because she should've known through the captain's wife that I wasn't going with them to East Prussia.

The time of my last military exercise had arrived. I had a beautiful pair of boots made to take home with me later. For the beginning of the maneuvers, I asked for a pair of military riding boots, since my own were genuine show-riding boots that couldn't be repaired. The sergeant refused. He told me to use my own. But he did send me on to the captain. I showed the captain the old military riding boots with the holes in them but he too told me I should use my own boots, no military boots for me anymore.

The following day during our riding exercises, I wore my old boots with the soles bound by twine and the spurs affixed with metal wire. Neither the captain nor the sergeant said anything and so I too kept quiet. And so it went for two days. In my mess kit, I had hidden a pair of slippers for after the exercises. During the third morning, while we were standing at Metz on a brigade terrain, there came a general riding on his horse. He asked the captain for a dispatch-rider. The captain searched around but the general shouted, "I want the right-flank man!" So the captain ordered, "Dielen come forward!" The general wanted to know whether I knew my instructions. I answered that I was the recruit instructor. "Follow me," was his response. Now the captain came toward me and told me I could go in the afternoon to the quartermaster at St. Marie to get another pair of boots.

I was quartered at a farm close to the general's quarters. After I took care of my horse, I asked the adjutant of the general whether I could get my other boots that were an hour ride away. The general, who was just walking by and happened to overhear my question, asked whether I had ever been arrested and if so, how many days? I replied in all honesty, "Not one day. Not even one hour, sir!" I received permission to get my boots, but on my return I had to immediately report back.

Fortuna, a beautiful and strong horse, ran like a deer to St. Marie where the quartermaster at first refused to give me a pair of boots. Only after a captain interceded was I able to get them. But the boots were not black and since I had to report back to the general, I asked to have them treated with black lacquer. After that was completed, I also asked for a pair of new spurs. That discharged another barrage of clamor and swearing. The captain, the sergeant, the shoemaker and the quartermaster were all agitated but I didn't say anything. Then they threw me a pair of spurs at my feet. Since I didn't have any tools, I asked the shoemaker to attach them for me. The group again began their railing until I mentioned I needed the new spurs because it was the general's orders and I had to report to him on my return. Then the sergeant was reprimanded and the captain, who'd up till then been fairly tolerable of me, said he'd personally make sure when I returned from the exercises I'd fade away in the stockade instead of going home. I put on my new boots, gave my *Fortuna* a piece of bread and in full gallop rode back to my quarters.

During the following days, I made sure I did everything as expected and looked where I could please the general or his adjutant. During one of the last days, I received a command to deliver an order at top speed to an infantry battalion. After carefully listening to the order, I raced in a gallop down the mountain. At the bottom of the mountain along a road lay our squadron. My captain stopped me and yelled I was abusing the horse and he was going to hold me accountable when I returned to the garrison. I showed him my orders and asked him, "What else could I do with the general's command 'at top speed'?" He turned around and mumbled, "You have your orders." My sweet horse *Fortuna,* who'd endured quite a lot from all the urgent orders of these last days, was again given the spur and off we went.

The major of the infantry signed off on my letter and on the way back, half way down the mountain, I noticed the artillery and also the infantry weren't doing what the general obviously had ordered. When I arrived back at the garrison, I reported my presence and told what I had seen. The general was able to correct the situation by issuing new orders so everything proceeded smoothly. After a couple of days, the army exercises came to an end and the major general Von Kähe went back to the city Metz; behind him followed his staff and myself.

After we arrived, the general called me in. He asked why it took me so long to deliver an order to the battalion. I told the general of the interruptions and the threats by the captain even though I didn't do anything wrong. The general gave me 10 German marks, a day's leave and suggested I do some sightseeing in the city Metz. On the day of my departure, I had to report again, wait a few minutes and was given a letter for the captain.

Then I returned to Diedenhofen where I arrived at 3 p.m. The reserves were just standing in attendance. I reported to my captain. "You're a day late, where've you been?" he said in a sour tone. I answered the general gave me 10 (Deutche) marks and told me to go sightseeing in Metz. But the captain only got angrier. He then summoned both the sergeant and the quartermaster and ordered them to take everything away from me; my horse, my equipment, everything. My comrades looked on in disbelief, but I was fed up. Before anyone could lay a finger on me, I took out the letter and presented it to the captain, which startled him. After he read the letter, he said that explains everything but still motioned for the sergeant to proceed in the taking of my horse. Then he said to me, "You're discharged." I then went with the quartermaster to exchange my uniform for a reserve uniform. Anyone who'd been in the service for 3 years would receive some second-rate equipment to take home with him. So I put the rags on as I left my good uniform behind. When I arrived at my room, my buddies were busy packing and clearing out. Out of my wardrobe trunk, I took my own brand-new boots, trousers and a hunter's jacket, together with a brand-new English shirt sent from home. I took my uhlanka (coat) and tossed it to the recruits. In the same manner, I threw my whole Royal-Kaiser reserve uniform to the new recruits. The other reservists would usually sell to the recruits for a few sous the things they didn't want to take along.

I started to dress when suddenly I felt a hand on my shoulder. It was the captain who said, "Well, well, Dielen, you didn't have that bad of a time here, did you?" I answered, "Well, the last few months haven't been all that great and I'll be glad to be done with all the harassment." Laughingly, he said, "Well, young fella, that happens in the best of quarters. Stop by my office in half an hour." My motto now was: get dressed real spiffy! I put on my taut-fitting blue trousers, my new riding boots with silver spurs, and the gray jacket trimmed with green borders and upright collar. I also put on my gray hunter's cap with the same green trim, complete with feather. Over one shoulder I had a reserve cap and over the other shoulder a new travel bag. With my pipe in my mouth, my quirt in my hand and a red silk handkerchief in my pocket, I stepped up to the captain's office. The sergeant and lieutenant Von Levegow were also there.

My knock on the door was answered with, "Enter." The captain sat with the back to the door, turned around, got up and saw me standing there. For quite a while he didn't say a word. He just looked at me from head to toe. Then he turned to the lieutenant, "My steward." Everyone laughed. Then, he took an envelope from his desk and said, "Inside this envelope is your evaluation. You, together with four others of the regiment, have received the qualification: *Outstanding*. Are you satisfied with this?" I told him that I'd never doubted his honesty and integrity; he laughed out loud as he offered me his hand while saying, "All forgiven and forgotten?" I then shook his hand. "If my assessment is correct, Dielen, we'll soon be seeing each other again. The French won't leave us alone." I replied, "When the war bugle sounds, I'll be the first one there." My fervor startled the captain and then all was quiet. Then Lt. Von Levegow asked me what was I now going to do at home. I told him my parents had done a lot for me these past 3 years during my absence and I had to make up for it with some butchering and sausage making. "Then you can send me a sausage someday," the lieutenant said with a smile. I promised to send him a sausage. I received my final discharge papers and could leave.

Back at the barracks in room 34, I had to tell my friends in full detail about my last meeting with the captain. The leftovers of the butter, bread and meat were divvied up and eaten and what remained from that was given to the recruits. We then quickly went past the gate and on toward the city. As soon as we came close to the city, we started to sing the beautiful song, "To my home, I wish to return." It was kermis time in Diedenhofen, so we had a real good time and luckily didn't have any problems. We were back in the barracks at 10 p.m. and after sharing a fine bottle of schnapps at the canteen, we went straight to bed.

The next morning, reveille was for us as late as 7 a.m. and after breakfast, we all went singing to the railroad station escorted by the music band. Once we got in the railroad car, the band played one more good-bye piece while in the meantime the captain came toward me to say, "Keep my offer in mind! Have a good trip!" With the sound of a final thunderous hurrah, the train set into motion and the many kitchen girls who'd come to say goodbye to their sweethearts were using their white handkerchiefs either to wave good-bye or to wipe away their tears. It'd take too long to describe all the adventures of the last 3 years, there were just too many. But the ones that left the deepest impression on me, I have written about. So, that's how my stay in Diedenhofen came to an end. I experienced a lot of distress, but also had a lot of good-times.

We arrived in Düsseldorf at night and were transported to the barracks. Whoever wanted to go on home could do so. Since there was still a train leaving for Krefeld; Lanken, Wänders, Eickelpasch and I decided to catch that last train and spend the night with family there. The following morning, we went first to Geldern to drop off Lanken at his home. Then, we three (remaining fellows) went on to Straelen. We arrived there at 4 p.m. Soon, we were surrounded by a lot of our old friends with whom we went to have a couple beers.

It was kermis time in Straelen, but Wänders and Eickelpasch wanted to go home first. So, I decided to do the same. After supper, I went to the ball for a while. It was the second to the last day of the festival, but somehow I wasn't in the right mood. I was too tired and had drunk too much beer to be a good dancer. Not to mention that my spurs were always getting entangled in the ladies' long skirts. Nonetheless, I hung around till 3 a.m. before I went to bed.

The upcoming day, my mother gave me a new outfit to be used at the 6 p.m. concert. My mother proudly took my arm as I was her escort to the concert. That evening there were many eyes focused on me, so I made sure to stay sober. After the concert, there was dancing again and I was able to whirl about the dance floor once more with girls I knew. Mother wanted to go home, so I brought her home. After dinner, I returned to the ball. That night I had quite a good time and the sun was about to rise by the time I returned home. Due to my sobriety, my friends and acquaintances remarked on my having mellowed out. Thus, the military phase of my life came to an end.

V. The Real World

The next day I had to get to work. My brother Louis and I surveyed the garden and land. We discussed what needed immediate attention. But during our noontime dinner, I got the impression Louis wanted me to make the decisions. I rejected that idea. After all, he was the oldest son and should be the one in charge. After discussing the issues, we came to a common understanding. Father and Mother were satisfied with our arrangement. That evening, Louis and I went out for a beer to conclude our agreement. And so we finally concurred Louis would take care of the business transactions and I'd work at home as butcher, sausage maker and with the assistance of a day laborer maintain the land or garden. So we were satisfied with our respective duties.

On Sunday, I'd usually visit the city because I didn't feel so attached to Straelen anymore. I gave the choral club another try and yes, sirree, the following year, while participating in a competition in Krefeld, we won the first prize. So while the choral club was progressing, we agreed there should be a drama club too, but that could only be staged in the Bürger lodge. There were a few obstacles in the way, but in the end, we prevailed.

Everything at the butcher shop worked out fine. Whenever there was a large wedding somewhere and our butcher shop had to supply the meats, I'd go and assist with the setting up of the tables, cooking, and slicing; free of charge. This information quickly got around and it wasn't long before no one would schedule a wedding unless Dielen was in charge of the supplies and the preparations. After the banquet was over, the party wouldn't let me leave unless I put on some play or recital. One thing led to another and it wasn't long before I was known in the whole area to the farmers and their marriageable daughters. You can imagine what that meant.

Our neighbor, Mathias Bocksteeger, married one of those rich farmer daughters from Grefrath. We, the Dielens, delivered all the meat supplies and I did all the preparations and cooking again. After this was all taken care of, I was invited to stay at the wedding. The father of the bride sat next to me throughout and wanted me to tell him all about my military experiences, because he too had served in the uhlanen, but in Düsseldorf. Meanwhile, Bella Hecker, the younger sister of the bride, joined her father and me. She wanted to know whether I wanted to sing a particular operetta. Luckily, I knew this operetta quite well. After a brief rehearsal, we got on stage. The audience liked it so much they wanted us to repeat our performance. During our repeat performance the band maneuvered itself onto the stage to prepare for the dancing phase of the day's events. I was then invited by Mr. Hecker and his daughter Bella to visit them later at the *Grierschen Boos,* a hamlet near Grefrath. Although I was more than happy to accept an invitation from these nice good-natured people, I made it clear it'd be "just in friendship."

The following evening, Bella and her brother came to pick me up, and whether I liked it or not, I had to join them in the ongoing wedding celebration. We celebrated the whole night through and by morning I had gained two new friends, the father and son. All the while, Bella was hanging on to me like glue.

A couple of months later, Straelen had its kermis and the Bocksteeger's invited the Hecker's. My memories of all the merrymaking that had occurred at their wedding banquet had by now somewhat faded. At the kermis ball, I ran right into Bella. Naturally, I danced with Bella a few times because the Bocksteegers were hoping I'd join them.

The following morning, as I passed by the Bocksteeger's home on my way to the garden, I encountered Bella and her brother. We talked about the fun we had at yesterday's ball and all the fun we had at their sister's wedding several months before. There was no doubt that Bella had fallen deeply in love with me.

One night, there was a burglary at the Hamer jewelry store in Straelen. Three suspicious fellows just so happened to draw my attention (that night) and ended up being detained as a result of my deposition. On the following Monday at 11 a.m., I had to appear in court at Kleve to give my testimony. In the meantime, I heard it was kermis time in Grefrath. So I planned going to Grefrath on Sunday and then Kempen and Kleve on Monday.

When I arrived at the *Grierschen Boos,* I learned Bella had found a boyfriend who just happened to be visiting *Grierschen Boos* too. Due to that, I just went in to say "hello" to Mr. Hecker and after having a glass of wine with him, said "good-bye" and left. I proceeded to a bakery named *van Boos.* The bakery had a café, a dancehall, and even rented bedrooms. The owner asked me who I was but I didn't tell him much. As the café and the dance hall slowly filled up with guests, the owner's daughter came in to see how everything was going. She half-laughingly whispered something to her father and left. Mr. van Boos then came toward me to say his son was an uhlan and was currently on furlough from Diedenhofen (Fr., Thionville). His son was temporarily out but would shortly return. It now started to get pretty crowded and a family asked whether I'd mind if they joined me at my table. They formally introduced themselves as Kamper-Busch, mill owners.

While I was talking to Mina Busch, the Hecker family walked into the café. Mina was 26 years old. She wasn't pretty but was friendly and pleasant to talk to. The Hecker family didn't pay any attention to me. I heard from others that Bella's boyfriend was an only son and heir to a large farm in St. Tönis. He was about 35 years old and as intelligent as a cow. Bella didn't particularly care for him but her father kept encouraging her, according to Mina Busch. Mina's brother remarked to Mina, "The fellow next to you is Bella's real love." "Oh, pardon me. Then you are *the* Hubert Dielen from Straelen! I wouldn't have recognized you." Due to the increasing jovial atmosphere, I invited my momentary companion, Mina, to dance with me. She was a light and good dancer. Bella danced with her brother because her boyfriend couldn't dance.

About a half an hour later came an uhlan laughingly and boisterously to our small table, "Corporal Dielen, how are you? I assume you'll be staying for a couple evenings at our kermis?" He said he knew me and knew all about the pranks I did in the military. I,

The Hecker estate in the year 1995

on the other hand, had no recollection of him. As a result of the uhlan's loud voice, farmer Hecker took notice. It started to dawn on me the party was a sort of social for the families of *Grierschen Boos.* I then told the café owner I didn't know this was a private get-together and thought it best I leave. The café owner's son overheard our conversation and asked the gathering for a moment of attention. He said, "Ladies and gentlemen, here sits a friend of mine and an uhlan too, Hubert Dielen from Straelen who inadvertently found himself at our gathering. He now realizes this is a private get-together and thinks he should leave. What shall we say?" "Let him stay!" they shouted. As I was still at the counter, I turned around and thanked them with a bow.

After a couple of dances, I was asked whether I'd be willing to put on a recital, but I declined. I didn't want to be the first. So Mina's brother stepped onto the table to great applause and proceeded to sing a song. Then, whether I liked it or not, it was my turn. I first did a skit of a farmer who had been to the big city. The people loved it and asked for more. The second skit was about an old retiree, who had a real appetite for young girls. The audience roared with laughter. Everyone wanted to have a toast with me. Even Bella, who had until now avoided me, came to me. Before I asked her for the next dance, I first asked for permission from her father and then her boyfriend. During the dance, Bella apologized for her father's attitude. I just laughed and shrugged it off. I said, "Bella, your father has misunderstood my intentions but that's okay." When the dance was over, Bella wanted to go outside with me. To avoid difficulties, I declined and returned her to her table. After another skit and a few more dances, it was the ladies' turn to choose their dance partners. Mina asked me first, but when she saw Bella coming, she sat down. Bella and I danced very calm and proper and Bella again asked me to go outside with her. She had something important to tell me. I steadfastly refused and escorted her back to her half-asleep and drunk boyfriend.

It was almost dawn when I had another dance with Bella. Before we all left, she said, "Hubert, when I commit to marry, it's till death do us part. A man like that one over there, I'll never marry, no matter how rich he may be and no matter how much my father may insist on it!" After the dance was over, I told her, "See you some other time in Straelen."

Someone had apparently paid my bill. Together with the family Busch, I left the party. The uhlan, who'd been drinking quite a bit, wanted to take Mina home, even though Mina wasn't too crazy about that idea. Nevertheless, he persisted and walked along with her. When we all arrived at the Busch's home, we were invited to have coffee with raisin bread and ham. There was plenty of everything. After that, I said good-bye to everyone and thanked them for the wonderful hospitality.

After arriving in Kempen, I looked first for a good restaurant where I could refresh myself a little. Then I bought a train ticket to Kleve. On the train, I found a good corner seat where I nestled myself in for a good nap until my arrival in Kleve. At the courthouse, I had to verify the identity of the three thief suspects I'd seen near the jewelry store in Straelen. I was called 3 more times at the court hearing before I was given a letter of thanks and compensated for my travel expenses.

In Straelen, everything seemed to be going fine. The drama club along with the choir was doing real well. The Bürger lodge didn't allow dancing during the kermis or carnival. So this gave our club the opportunity to provide comedies on Sunday evenings. I arranged with the leadership of a walking band, *The Harmony of Venlo,* a regular exchange of concerts and choirs between Venlo and Straelen. One day, after a successful performance in Venlo, I went with two friends to the Engels café at the Maaspoort to have a beer. We knew Mr. Engels had four marriageable daughters, but the daughters were a disappointment to my boisterous friends because the daughters were not particularly friendly and were kind of sour, with the exception of the oldest one who was at least somewhat pleasant. I took it as a challenge. It's not that big of a deal to approach a friendly young woman. Anybody can do that. But to charm a reserved, sour young woman, that's more fun. My friend, Sjeng (that's another way of spelling John) Erprath, agreed. So every time Sjeng and I were in Venlo, we went to the Engels café at the Maaspoort to have a beer. Mr. Engels knew the Dielens quite well and often shopped at our Dielen business to buy bacon. Mr. Engels was always cordial and in the best of spirits when Sjeng and I would spend a lot of money in addition to buying him a couple mugs of beer. Slowly, I became well-known in Venlo.

At the hotel *de Poort of Kleef* (transl. the Kleve Gateway) in Venlo lived a German named Piet Stommel. He was a reserve officer who had to regularly report in Straelen. He also soon became one of my buddies. It wasn't long before he was referring quite a few customers to our business. He even promoted our well-known *frankfürter-knakwürst.* Before long, I was making frequent deliveries to Venlo. When I had to make a Saturday delivery to Herongen, I would continue on to Venlo with our well-known sausages via Niederdorf. As a result, I regularly visited the Engels family with some knackwürst. I was soon accepted as a household friend. I no longer saw (the people from) *Grierschen Boos* anymore. I only heard Bella wasn't allowed to come to Straelen because she now had a rich farmer's son as a steady boyfriend. It made no difference to me because I was still getting my recreation and practically every Sunday I was out of town.

It was Kermis again and on Sunday evening our drama club performed the play "Yearning Ladies" at the lodge. The character I had to play was Bruno Waldmann, which had the first appearance on the stage. The mayor, Hermkes, was seated in the prompter's booth and everything was ready. The prompter was ready and I knew my part pretty well and I let the prompter know I was ready to go. Then I looked up at the audience, and who do I see there sitting at the first table right in front of the stage? That's right. I see the Bocksteeger family. Bella was focused on me with big eyes. I had to get a hold of myself and swallowed a couple times. Then I began, "In my deepest feelings injured, like a beggar thrown out! Yeah, Oscar, you were right to warn me of such a step. But then, what does one not do for the sake of love?" Only my voice was heard above the eerie silence of the audience. The mayor nodded at me with an understanding smile. The part fitted my situation with Bella pretty well. No wonder I played the character perfectly. At least that was the opinion of the public, who told me so during the intermission. They could understand very well how I was personally affected. When I walked by the Bocksteeger's table during intermission, Mr. Hecker stood up and congratulated me on my acting. Then Bella also stood up and said, "Father, if you're allowed to talk to Hubert, you can hardly forbid me from doing so." So she also congratulated me and then offered me a glass of wine. I declined explaining I couldn't accept any drinks (since I still had to finish the performance). I then looked for the table of my own family. Then the bell rang and the players went backstage. From then on, the play continued smoothly. When a moving verse was spoken, the audience would spontaneously applaud. Quite a few of the ladies in the audience were aware of the coincidental similarity between Bella and myself in the real world and thus applauded all the more. When the play was over, my family and I lingered around a while before going home. On my way out of the lodge, Bella whispered to me she'd be at the garden in the morning at 9 a.m.

I had a hard time falling asleep that night. At 9 a.m., I used the back door to go to the garden. That way I wouldn't have to pass by the Bocksteeger's house. Bella showed up within 10 minutes with her younger sister. I asked Bella to join me in the garden house where I offered her a glass of a mild liqueur. The younger sister all of a sudden became very interested in the flowers of the garden. Bella told me what her family had said last night. Her father continued to pressure her to marry the rich farmer from St. Tönis. Her mother, sister and brother sided with Bella, who wasn't so eager. Bella gripped both of my hands and said, "I'd rather be dead than marry that man and if you don't want me, I'll enter the convent. I swear to God." I told her I wasn't interested in marrying anyone for the time being but we could still be friends. Besides, I had no intentions of marrying a farm girl. My wife would have to be a businesswoman. At that, she started to cry, "I'm only 24 years old and still young enough to learn new things in addition to being a determined person." I gave her a comforting smile and said, "Oh, that's all right, we'll talk about this some other time." The younger sister of Bella returned to the garden house with a bouquet of roses in her hand and asked, "Should I be congratulating you two? We have to go home now because the new guests may have arrived! Hubert, Bella's suitor (Gerard) is coming today. Maybe, I should offer him this bouquet of roses from your garden. Wouldn't that be nice? Boy, will that fool be surprised!" Laughing and giggling like schoolgirls, the two skipped out of the garden. I stayed behind for another hour in the garden house to think things over. After that, I got up and went home. This time, I took the regular route. As I passed the Bocksteeger's family home, I noticed several people looking up but I didn't stare back. I first had a nip to drink (at a local café) before I went home for my noontime dinner. Louis, who didn't go out much, was already home. We didn't wait for dad, since he was elsewhere.

Although a kermis was going on, my family didn't have many guests at home. In the evening, we'd all go to the kermis. Later that evening, we encountered the Bocksteeger family. Lately, I had been quite friendly with the Theuniszen family and would often have a dance with their daughter Stiena (that's short for Christina). She too was quite fond of me. When I asked Bella for a dance, she told me their visit to our garden didn't fall on good ground (meaning her parents weren't happy) and she had to promptly go home in the morning but she'd surely write me.

Straelen's kermis winded down and then a while later came an invitation from Mina Busch to go to their kermis in Grefrath. Unfortunately, I couldn't accept. Lately, my beloved mother had again fallen ill and shortly thereafter went to her reward July 25, 1880. We had to hire a housekeeper to do all the household chores my late mother once did. It didn't take very long before many items in the house disappeared. The housekeeper got fired and we had to look for a replacement. My dad, to no avail, urged Louis to start looking for a wife. Then Father suggested I should marry Stiena Theuniszen and still live at home. I balked at that idea. Louis was supposed to live in the family home. After all, he was the oldest. I could always find another place to live somewhere in this world. My dad and Louis thought my viewpoint simply grand and shook my hand. Around this time, my brother Philip came home because he was quite ill. He tried to help a little but couldn't do much.

A letter arrived from Bella, she asked me to meet her at a relative's house in Wankum. I accepted and punctually arrived in Wankum. Bella was alone waiting in the visitor's parlor. She began a very intricate conversation. Bella told me that she had done exactly what I had advised her to do and wouldn't go against her father's plans and didn't argue with him. As a result, the wedding announcement was going to occur the coming Sunday and in two months the wedding ceremony would take place. This revelation really put me on the spot. Bella asked me again, what she'd have to do to become a good businesswoman. I told her, "First work as a store clerk in Venlo for 6 months and then another 6 months as a cook in *de Poort van Kleef* restaurant."

Bella's face lit up while she grabbed both of my hands saying, "Great, then that's what I'll do." With this, she felt her problem resolved. I asked her what else she wanted. I seemed to have gotten myself into a sort of conspiracy. What was she going to do next and what was I supposed to do? Bella had it all figured out. "Just make sure I have a place to stay when I arrive in Venlo." Bella and I said good-bye to the van Laar family who were curious as to when Bella's wedding was going to take place. "Not yet this year," answered Bella with a smile. It was getting dark and so Bella and I left. I then escorted Bella through the woods toward her house nearby. "I'll write you soon, Hubert," she said as she waved good-bye to me.

The days that followed were full of anxiety for me. In Venlo, I quickly took care of business. But on Monday, I received a letter in the mail. Bella wrote that in fourteen days, she'd be visiting her pastor and that the following Sunday would be the first banns. Then there'd be a reception and I could go to give my congratulations but that didn't mean she was actually going to follow through. I could be sure of that as long as she was alive! "When you write me, send the letters to my friend Wilhelm. He also takes care of the letters I send to you. I have my plan all set. I'll gather my trousseau and have it sent to Venlo. Then on the last Sunday, I'll send a letter to Gerard saying I won't marry him. Likewise, I'll inform the pastor of my changed plan. I won't marry Gerard under any circumstances! My friend Wilhelm will keep you posted. He helps me where he can. My brother and sister also help but don't know about my plan. Until later, Bella."

Julius Lankes, a Straelen horse-trader and former uhlan of whom I sometimes assisted with his trade, readily gave me a horse to ride whenever I so desired. In his stable was a fine Dutch artillery horse that had four strong legs. The stableman groomed the horse and saddled it. I put on my riding breeches and boots and in less than 30 minutes, I was on my way. It was custom in Grefrath that on the Sunday of the first banns, friends and acquaintances would come to the house to offer their congratulations and so I decided to go as well. Before I reached Wankum, people were already asking me whether I was on my way to *the Boos* for the celebrations but I didn't say anything and rode on.

As I was coming out of the woods with *the Heckerhof* in sight, I could barely believe what I saw. Bella was leaning out of an upper window waving a white handkerchief at me! The swing-gate of the farmyard was closed and there were people standing at the door. After giving the horse a few switches with the quirt and a light nudge with the spurs, the horse was galloping forward, one more nudge and the horse jumped while in gallop over the swing-gate into the farmyard. I dismounted the horse in front of the door where farmer Hecker and Wilhelm stood. I shook Hecker's hand, congratulated him and gave him a (congratulatory banns) card for the fiancée. The old fella stood stunned.

I remounted my horse and departed in the same way I came, that is, in full gallop over the swing- gate. At the Mulders, I received a cup of coffee. I didn't say anything about what had just occurred and felt pretty smug. After a half-hour, I again passed by the Heckers. Although the swing-gates were now open, I continued on my way to Wankum as I waved my handkerchief (at the Heckers). There I had myself a beer at van Straaten and then trotted back to Straelen. There, the gossip of the whole affair had already taken on huge dimensions.

On the agreed upon Sunday, I was already in Venlo at *de Poort van Kleef.* I was just about to go to the railroad station when Wilhelm arrived. He said that while Bella went to the High Mass in the village 20 minutes from the *Heckerhof,* her family discovered that all of Bella's clothes were missing. Bella's father tracked her down with his horse and carriage and compelled her to return home. She had already talked to the pastor to retract the banns. I wanted to go with Wilhelm on the very next train back to Grefrath but Wilhelm strongly advised against that idea. He said, "Hecker is a hunter and if you go near him, he'll surely shoot you in the head because last Sunday he already wanted to shoot you off the horse! That's the reason I stayed near him. So that had Hecker gone for his gun, I would've been able to interfere!"

Wilhelm went back to Grefrath and I went to Straelen. He promised to keep me informed. It was understandable that I was quite upset. I went early to bed that evening, which really surprised Father and Louis. Louis thought something had already occurred. Lately, there had been a lot of rumors in the cafés about me, Hubert. Early the following morning, Wilhelm came to see me. He told me he had spoken to Bella that night via her bedroom window and she told him to tell me not to visit her under any circumstances, for that would certainly result in a casualty. I'd hear from her later. Wilhelm was a 26-year-old rich farmer's son and neighbor of Bella. Wilhelm and Bella had been childhood playmates before they had walked to school together in all sorts of weather. Not too long after this, he got engaged to his half-cousin Gertrud Steckelbrock from Wankum.

It didn't take very long before the whole story about Bella and myself was known all over Straelen with the obligatory fabrications, of course. Since we were still in mourning because of the death of my mother, I didn't go out much. On Sunday, I'd go either to Venlo or to Geldern and sometimes to Nieukerk but at night I'd stay home.

In the meantime, I'd told Louis the whole story about Bella. At first he laughed, but after thinking it over a little more, he pensively shook his head. After a couple of weeks, I heard Bella went to St. Tönis with her father to apologize to Gerard (her former fiancé). Gerard didn't accept the apology.

Days before, I received a letter from Bella in which she apologized to me for all the trouble she caused and she'd defer to her father's wishes. I immediately wrote her back and wished her the best, not knowing that Gerard in St. Tönis had already shown her the door. Gerard's rejection of Bella saddened me. She was after all a cheerful and friendly girl with good character.

There were a lot of bad stories about me going around. It even got to the point where my father, who was rather benevolent, was blamed for not paying more attention to the goings on of his son, the uhlaan. My father returned drunk and agitated (from a pub, where he usually chatted with locals). He would have hit me with a chair had it not been for Louis, who placed himself between Father and me. Louis wanted me to immediately explain to Father what really had happened but I said, "Tomorrow." So Louis told what really happened. Father then calmed down and we all went to sleep. The following morning, I asked Father what was everyone saying about me but he didn't want to repeat it.

So Louis began to repeat the stories, but I stopped him and said to Father that if he didn't repeat the stories to me, I'd pack up my belongings and leave for good to America or so. Louis now insisted Father repeat the stories that were being told about me. Father began by saying how the family's good name and reputation came first, even before the reputation of some rich farmer's daughter. I agreed with him wholeheartedly. When he finished repeating the stories, I told him most of the stories were untrue and my part in the real story was done merely as a lark. I then told him the full story and how it all began. I asked him not to feel bad about it anymore and not to talk about it either. As for the good name of Dielen, I'll worry about that myself. Father was deeply touched by my candor. Thereafter, the gossip in Straelen about Hubert Dielen gradually faded away.

The joys of bachelorhood eventually came to an end. But on one particular morning, I was invited to the Bocksteeger's. I arrived at 11 a.m. I was cordially received in the salon by Mr. Bocksteeger, who was still a good friend of mine, and offered a drink and cigar. But what I wanted to know was what was the honor that brought about this invitation. Bocksteeger then called in farmer Hecker. I greeted Hecker with reserve. Bocksteeger asked farmer Hecker to explain why he, Hecker, thought it necessary to talk to Hubert Dielen. Hecker said Bocksteeger was well aware of all the rumors about Bella and Hubert Dielen that were making the rounds in Straelen and Grefrath. I told Hecker I didn't know all the rumors and asked him to inform me. Bocksteeger then slowly began with Hecker's assistance to repeat all kinds of rumors saving the worst rumor for last. I stood up and said, "For her and God I swear that Bella, as far as I'm concerned, is as pure and innocent as an angel." That hit like lightening from heaven. I further said that as far as all the other rumors (about me) were concerned, they didn't interest me and I could care less. I must say my friendship had been poorly reciprocated and could even have cost me my life. They sat as if they were glued to their chairs. After a moment of silence, farmer Hecker said, "If you had come to me in a proper manner and asked me for my daughter's hand, I certainly wouldn't have objected and you still can do so." I jumped up and said, "Did you think you could just shoot at me as if I were a hare in a burrow?" Hecker got upset and wanted to leave but Bocksteeger, who knew how to calm down a person, told him to sit down and offered Hecker and me a drink. Bocksteeger began to recall all the good times he had had; like at the kermis, the first time he had met his wife, his own wedding and so forth. His stories altered our moods and the atmosphere became more congenial. A couple of days earlier, I had heard Bella was dating a Mr. Schmitt of Grefrath. So when Bocksteeger intimated he and

I should visit the *Grierschen Boos* one more time, I quipped, "Let's first see what happens in the next nine months." So that's how that came to an end. As we got up and started to leave, Mr. Bocksteeger gave me a package of knickknacks from Bella.

When I got home, the noontime dinner was already on the table. Father and Louis knew I was at the Bocksteeger's and had waited with eating until I returned. I told them the gist of the story then added Bocksteeger and I planned on going to the *Grierschen Boos* this coming Sunday to witness the announcement of the real wedding of Bella. Father laughed wholeheartedly and turned to Louis, "Now you better look for such a rich bride!" I had to ask Father, "Where will the honor of the Dielen name then end up?" That ended that subject for a while. But after dinner, Father again asked, "Do I understand you correctly? You're not interested in that girl?" "That's right Father, I'm not interested in her." They laughed and then we all went to the garden. I never saw Bella anymore and later heard that she actually married that Schmitt fellow. Bocksteeger and I remained friends.

VI. Mina (1st wife)

One day, Sjeng Erprath, who had to do some buying in Venlo, asked me to join him on a business errand. Anticipating the purchases, we took a horse with wagon. In Venlo, we hitched the horse to a post and then went about our business. After we finished, Sjeng, myself, and some others went to the Engels café at the Maaspoort for a refreshing glass of cold beer. The Engels family had just finished butchering and was almost finished with their sausage making in the cellar. Sjeng hung around the kitchen where two of the sour faced sisters were. Nonetheless, the sisters did offer him a warm piece of sausage. I too was offered some but declined and went to the cellar where Mr. Engels was working. Mina came downstairs with some beer from the kitchen for her father and me. Sjeng was a jovial type of guy who was trying to entertain the two somber sisters in the kitchen and then later in the living room. Mina was a nice, hard working girl, who had lost her mother when she was only 14 years old. She was the oldest daughter and the one with the pleasant disposition. Mina later joined Sjeng and me in the living room where Mina and I talked to our hearts content. While we were talking, I suddenly blurted out that it'd be a shame if she didn't become a butcher's wife. Mina replied, "I would if I could find a good butcher." "Well, why shouldn't you?" Sjeng promptly said as he took Mina and me by the arm and placed us arm in arm in front of the mirror. "Don't you think they make a fine couple?" Sjeng remarked to Mr. Engels who was just coming up the stairs from the cellar. Mr. Engels poured himself some beer and then partook in our congenial conversation. Then

Sjeng and I said our good-byes to the Engels family, loaded our wagon and returned to Straelen.

As soon as Sjeng and I were outside the city, Sjeng said, "Mina sure is a nice girl and the moment she shows an interest in me, I'll take her. You can be sure of that!" The following Sunday, on the 4 p.m. train, arrived Mina's youngest brother Tinus (the Dutch name for Martin), who came to Straelen to visit me. He was a common visitor amongst my family so he was promptly offered coffee. When I suggested we visit Sjeng Erprath, he rejected that idea. He first wanted to converse with me. That got my curiosity going. He then told me Mina wanted to invite me over the following Sunday without Sjeng and I had to promise to come without Sjeng. I had nothing planned the following Sunday, so I shook Tinus's hand with the agreement to come without Sjeng. We then sauntered through Straelen stopping here and there for a beer until Tinus's train arrived for Venlo.

The following Sunday, I had had a few beers too many and was a little inebriated. I had fallen asleep on the couch until the housekeeper woke me up. I ran to the railroad station and was just able to catch the train at the last minute. After I had arrived in Venlo, I noticed in the reflection of a shop window that I wasn't wearing a necktie. Lucky for me, on the corner of Klaasstraat (St. Nicholas Street) and Vleesstraat (Meat Street) lived an old acquaintance of my family who sold neckties. Her name was Mieke Joosten. Mieke sold me a necktie and tied it on for me. She asked whether it was true I was engaged to one of the Engels girls. She also mentioned how there wasn't much money in that family and that she, Mieke Joosten, was much better financially situated and wouldn't mind being married to a good butcher. I smiled at her as I said good-bye and went on my way toward the Maaspoort. Mina stood at the entrance with a happy face and took me by the arm and led me inside.

Inside, there was a lot going on. Mina disclosed what had occurred and insisted I not take offense. She apparently didn't have much choice at the time. You see, a couple days after our previous get-together, two gentlemen from Eindhoven had visited the Engels home. Her sister, Hubertina, was engaged to one of the men. The two men continued to challenge Mina as to her not having a fiancé. Mina finally told the fellows she did indeed have a boyfriend, a real nice one, but it was still a secret. The men made a bet of 5 bottles of wine for Mina to show them her boyfriend this Sunday when they had to return to Venlo to sketch a picture of the Wettelings reception hall. "That's the reason Hubert, and please don't be annoyed, I called on you. After all, didn't your friend Sjeng remark about us making a nice couple?" Perhaps because I didn't immediately say anything, Mina then became real shy and blushed. But I too found it all quite amusing and said, "Very well. Let's go forward, then."

Mina took me by the arm, knocked on the living room door and we entered, whereupon Mina introduced me as her fiancé. I made a bow and said, "My name is Hubert Dielen." Everybody got up, applauded and shook my hand. Five bottles of wine were forfeited to Mina and me. Mr. Engels was quite surprised. He didn't get it. At the table, two seats had been reserved for Mina and me and so we all sat down to a good meal and coffee. When the 2 gentlemen, Mr. van Dijk and Mr. Booms, had to leave to (sketch a picture of) the Wettelings reception hall, they invited Tinus, Mina and me along. Stepping outside, Mr. van Dijk suggested I offer Mina my arm and then we all went laughingly to the Wettelings. At Wettelings, another two bottles of wine were made available of which we all drank while the two fellows from Eindhoven went about their work.

The evening continued on in much enjoyment and the two gentlemen caught the last train to Eindhoven. But I, something that had never happened to me, missed my train. Now, Tinus in the middle of the night had to drive me in the horse-and-buggy all the way to Straelen. When my father heard the story (of my winning the bet), he was quite happy for us, but nonetheless hoped we three brothers would stick together and not get married. But I had told him many a time I had had enough of Straelen, and my knowledge of Venlo in addition to my relationship with the Engels family had me set on marrying Mina.

So I went to Venlo and asked Mr. Engels what he knew about the property Groenmarkt (tr., Vegetable Market) 560, which was available. He told me the house belonged to Mr. Wolters, who had purchased it at a bankruptcy proceeding. Mr. Engels didn't mind going with me to look the house over since he felt it was still in good condition. He could then also speak with Wolters, whom he knew very well. So we went together to the market (Groenmarkt), where a very friendly Mr. Wolters showed us the entire property.

Wolters wanted to rent the property out for 600 guilders (most likely per year, at least). I told him I'd make my decision and get back to him that same day. Back at the Engelses', Mina was already waiting for us curious as to the outcome. I told Mr. Engels I'd rent the property under one condition. Mina then made an effort to leave the room, but I sat in front of the door and didn't move aside. Mr. Engels wanted to know what the condition was because he didn't feel the rent was his problem, nor would he be a guarantor. I said it was much more important than that. I'd only rent the house provided he allow me to marry his daughter, Mina. I then offered Mina my arm as we stood in front of Mr. Engels. It was very difficult for the old fella to suddenly part with his Mina who had worked so hard for all those years keeping house. After I promised to take good care of Mina, he gave his approval. Her brothers and sisters were called in and with a good glass (meaning a couple glasses of wine or beer) was the engagement announced. After that, I went to Wolters, rented the house and returned to Straelen.

Mina and Tinus came to Straelen on Sunday to invite my father to go the following Sunday to Venlo. So Father did and when he returned he appeared quite content. He promised the Engelses he'd return the following Sunday. So when that Sunday came and after we had had our noonday meal and Father his nap, I had the horse and carriage all set to go. Assisting Father into the carriage took a while, but once done it wasn't long before we were soon in Venlo. I stopped for a moment at the Groenmarkt and pointed out to Father the house I was going to rent. He looked stunned and asked, "What are you planning?" I replied, "You're from Venlo and I'm returning to Venlo." Father seemed to accept my reply quite well and soon we arrived via the Steenstraat (Stone Street) at the Engelses' front door. Mina helped my father get out (of the carriage) and once we were inside (the house), I introduced her to my father as *my fiancée.* Father looked a little surprised but nonetheless shook Mr. Engels's hand and also Mina's hand. During and after our afternoon coffee, we deliberated future arrangements. I then said I was already renting the house. Father thought I was a little too hasty but Mr. Engels didn't think so.

September 8th being the Blessed Virgin Mary day, the subsequent Sunday was when the kermis in Straelen began. My sister, Drika, was visiting from Borbeck (today it's a suburb of Essen, Germany) to partake in the kermis. Mina arrived on Saturday night. At 3 a.m. on Sunday, Drika, Mina, and I awoke so as to go by foot on a pilgrimage to Kevelaer. After praying very hard and going to Mass, we had ourselves a hearty breakfast. We returned to Straelen via Walbeck. We arrived in Straelen just when the noontime dinner was put on the table, which was a good thing because we were pretty hungry.

Drika and Mina spent the rest of the Sunday preparing the furnishings for the store, while I busied myself with the carpenters who were making the furniture. Mina and I didn't go to the kermis because we were so busy preparing for our wedding on October 2nd, (1883). On Wednesday, Mina and I went to Venlo to get the workmen started in our new house. Everything went smoothly and by October 1st, the store was ready.

[Hubert's announcement for the opening of his new store has been translated thus.]

☞ The undersigned makes it known that he has established himself at the following address this coming *Saturday, October 6th.* His

Pork Store

with the accustomed complementary products will be located at the house, Groenmarkt No. 560, and open for business. He courteously wishes to present himself at your service.

H. Dielen-Engels

For Rent:

A very nice, above-store, living-quarters.

(From the Venlo Weekly of September 29th, 1883.)

Wilhelmina Engels 11/12/1852-1/10/1890

Six weeks earlier, a young man from Venlo, named George Haanen, had opened a butcher shop right next to mine. A year earlier, across from my store, a Jozef Faldera also opened a store. They both came from financially sound families. But I wasn't too concerned, because I figured I'd be able to get some working capital from Mr. Engels. I also expected to get some working capital from my father, but not much. Thus, I wasn't too worried about my competition. Before finally moving to Venlo on Monday, October 1st, 1883, I had to finish up on some duties I still had at my home in Straelen. From my father, I received 600 German marks working capital. "Hubert, this money is for you. You have earned it." He also advised me to use this money wisely. Please don't ask for more, because I can't afford giving more. I tried to hide my deep disappointment.

Meanwhile, in Venlo at the Markt (No. 560) and at the Engelses' place in Maaspoort, everyone was busy with the preparations for the upcoming wedding. I gave Mina the 600 marks and the key to the business and told her to take good care of both. Mina felt honored by my entrusting her with such responsibility. All the relatives and friends of the Engels family were invited to the wedding. There were about 80 people present on Tuesday. From Straelen were invited my father, Drika, Louis and Fien Peters. Fien was a good old friend of the Dielen family and was the bridesmaid at the wedding.

On the Sunday before the wedding, I was told by a friend, Jos Linszen, who was the secretary at the Venlo town hall, that the wedding couldn't proceed as scheduled Tuesday because the wedding announcement hadn't been on display (for 14 days), as was the local custom, on the (town hall's outdoor) bulletin board (which was encased behind a glass panel). So (on Monday) I went to the registrar's office. Jansen, the registrar, averred that he had forwarded all the paperwork on time. I then went to Jos's office and for 5 guilders he made a new announcement paper, which he then placed half hidden behind the others (on the bulletin board). (Back inside the town hall,) The mayor maintained he hadn't seen the announcement paper. So we went outside (to look at the bulletin board) and sure enough, there it was. "Yep," said Jos (winking), "there it has been for 14 days." The mayor chuckled in disbelief and played along by signing the papers as of then. Back at the registrar's office, Jansen asked how everything turned out.

When I told him we found the announcement, he smiled primly. I later heard that Jansen and the Engels were not on speaking terms. That explains the displaced papers.

Another incident that almost sidetracked the scheduled wedding (and civilian life) occurred earlier in that year when I received a summons from Münster to report for an 8–week training to be a squad leader. To be summoned for such a position was quite an honor for a respectable German (like me). Now came the question, how do I get out of this? After all, I wouldn't have been able to return from this training course until mid-October (hence delaying and perhaps thwarting my original intentions in life). I had only one recourse, and that was to ask for an urgent emigration form from the ministry in Düsseldorf, which would then remove me from my military obligations. Düsseldorf forwarded my forms to the mayor in Straelen for further information. The mayor, who happened to be a good friend of the Dielen's, asked me whether it was true I was emigrating to America because he heard I was only moving to Venlo. I told him the whole Engels family was emigrating to America and I was going with them, of course. If the forms weren't prepared in time, please have them forwarded as soon as possible to my new address. If the ministry in Düsseldorf rejects my request, I'll already have been long gone. "Well then, I might as well sign these papers," said the mayor. He signed the papers and told the courier to go ahead and mail them. Once the courier had left, I showed him my military summons. He snickered and said, "I knew there was more to it."

Now, back to Venlo; all went well at the wedding but unfortunately for me, I had to go to work the following morning since Mr. Engels wanted me to open my new business that week. Something I wasn't too enthusiastic about. I also had to buy cattle from a new and unfamiliar location. So Tinus Beerden, the fiancé of Mina's sister, went with me. Since Tinus had known the Engels family for many years and also appreciated my industriousness, he helped me where he could. Tinus had also loaned me some beef and a good well-fed calf that weighed about a 100 kilograms. On our way there, he mentioned how everyone in the Engels family (with the exception of Mina) was unusually selfish and Mina wasn't loved by her siblings even though she was the best one. The reason being that after the death of Mina's mother, Mina was directed by her father to be a strict disciplinarian, a sort of second mother. Mr. Engels loved his oldest daughter very much, which only exacerbated the hate and jealousy from her siblings.

As soon as Tinus and I arrived at the farm of farmer Heiting, in Baarlo, we bought 2 pigs. When we came home at 3 p.m. that day (Wednesday), Mina already had the dinner prepared. By 6 p.m., the entire Engels family was at the house dining on the leftovers and finishing off the drinks. It wasn't until midnight, after an evening of good cheer with everyone, that Mina and I were alone.

On Thursday morning, the 2 pigs arrived. It still bothered me that the premiere opening of the new business was occurring during the first week of my marriage, but I had to accept the fact that the opening day had already been decided. An advertisement was placed in the local daily paper to further announce the upcoming premiere opening. The pigs were butchered somewhere else and the carcasses returned (from which was prepared the meats that were then put behind their respective display windows). I was then hungry and looked around for something. Mina said with a sad face that there wasn't anything leftover from last night. "Well then," I said, "I'll find something." So I went to the sausage kitchen, cut a piece of meat from one of the two still warm pigs, fried the meat with some potatoes in a pan, brewed a pot of coffee and we had ourselves a fine noontime dinner.

On Friday, I had to start working before daylight. A good friend, Fritz Franszen from Geldern who planned on helping me, canceled out. But Louis brought a large portion of saveloy and Mina helped wherever she could. Before evening, we had enough sausages completed to fill all of the display platters. Tinus Beerden took half of a calf carcass and in return brought a fine portion of beef.

On opening day, Mina stood ready in the store with her bleached white apron. We right away had a lot of customers. I knew we had the biggest, busiest, pork store in Venlo. That Saturday, I hired an apprentice from Straelen to run errands and other stuff. The premiere opening was a success. Mina was in high spirits and was selling one item after another. We were so busy we didn't have time to eat.

Saturday afternoon, the newspaper arrived and what do I see? There was printed a good-bye poem for Hubert Dielen from the Straelen choir club *Concordia.* I later found out the poem had

[Here is the poem penned by the chairman of the choir, mayor Hermkes. The poem, written in German, has been translated thus.]

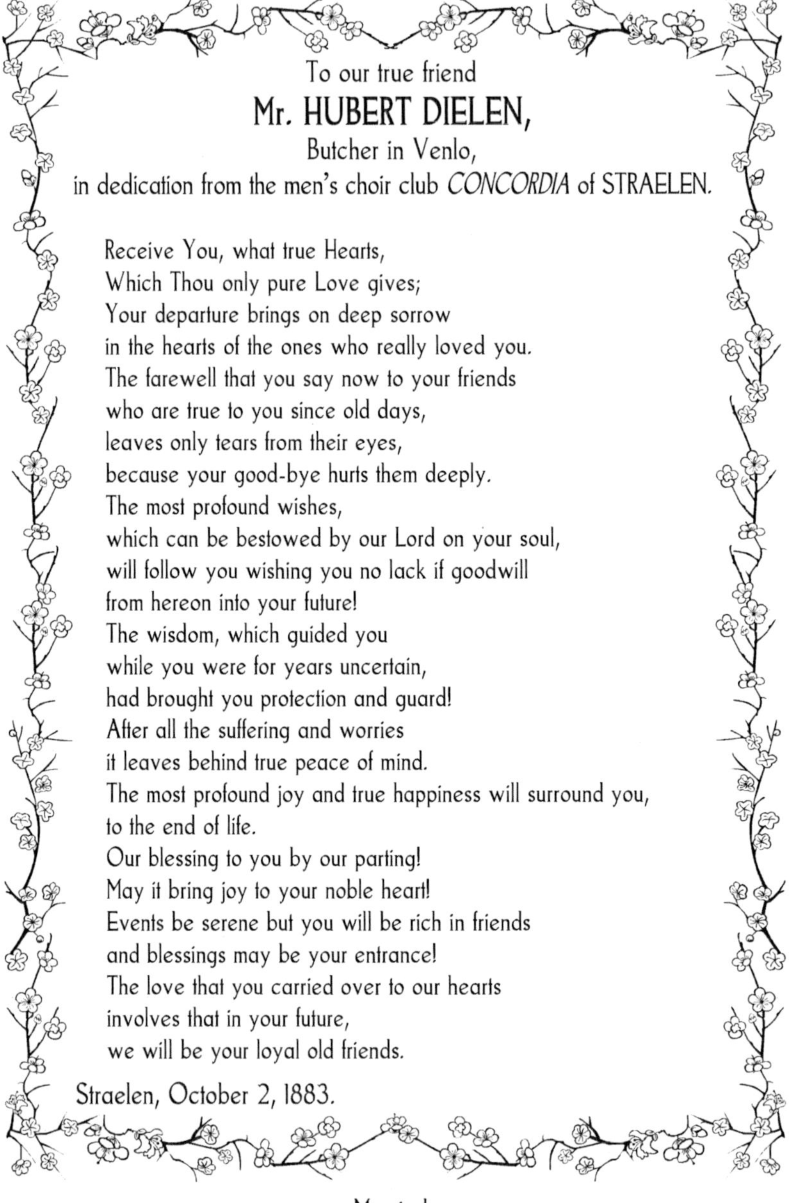

To our true friend

Mr. HUBERT DIELEN,

Butcher in Venlo,

in dedication from the men's choir club *CONCORDIA* of STRAELEN.

Receive You, what true Hearts,
Which Thou only pure Love gives;
Your departure brings on deep sorrow
in the hearts of the ones who really loved you.
The farewell that you say now to your friends
who are true to you since old days,
leaves only tears from their eyes,
because your good-bye hurts them deeply.
The most profound wishes,
which can be bestowed by our Lord on your soul,
will follow you wishing you no lack if goodwill
from hereon into your future!
The wisdom, which guided you
while you were for years uncertain,
had brought you protection and guard!
After all the suffering and worries
it leaves behind true peace of mind.
The most profound joy and true happiness will surround you,
to the end of life.
Our blessing to you by our parting!
May it bring joy to your noble heart!
Events be serene but you will be rich in friends
and blessings may be your entrance!
The love that you carried over to our hearts
involves that in your future,
we will be your loyal old friends.

Straelen, October 2, 1883.

Married:

H. Th. DIELEN

and

Wilhelmina ENGELS

Venlo, October 2, 1883 Reception Groenmarkt

From the Venlo Weekly Magazine of October 6, 1883

been penned by the choir chairman, mayor Hermkes. What splendid free publicity that was! In the late afternoon, I received a message that the *Concordia* choir would be coming Sunday at 4 p.m. By Saturday night, all the merchandise had been sold with the exception of the suckling pig I had prepared and then roasted in the Engelses' oven before putting it in our display window. At first, we were going to sell the pig, but then I thought we better wait until I found out what the choir needed.

On Sunday afternoon, it was pretty crowded on the street. Mina and I didn't pay any attention to the crowd since we were so busy preparing a reception for the *Concordia* choir. But my (choir) friends already stood in front of the door and in full force they sang: "To honor a parting friend, we as friends come together today and restrain with difficulty our sorrow of his parting but remember with pleasure the good-times." There were about a 100 onlookers in front of the house. After the first verse, Mina and I walked arm in arm to the front door. A thunderous "hurrah," occurred. Mina's tears flowed like a river and I had a hard time keeping my own eyes dry. The second and the third verse followed. I stepped forward and thanked the gentlemen for their considerate thoughtfulness, their presentation being something of which Mina and I wouldn't easily forget. Mina and I invited the 34 choir members inside but the public wanted an encore. So the choir sang one more song. The choir felt relieved when they finally came into the store. Mina and I arranged seating in the front room and partly in the store so everyone was able to find a seat. We still had enough beers and cigars left over from the wedding celebrations.

While everyone was enjoying their refreshments, I asked everyone for a moment of their attention. I was pretty nervous and began with a thank-you for the beautiful farewell poem they placed in the Venlo Weekly and the honor they gave me by their presence. I mentioned how the other butchers in Venlo were apprehensive of the upcoming competition and were doing everything they could to make my life sour. But the people of Venlo, after reading such a farewell poem and listening to such an acclamation (as given that afternoon), have surely become aware (of the extent I've been appreciated back in Straelen). I no longer worried about my acceptance (in Venlo). I candidly admitted I had many a sleepless night (due to such apprehension), but was now at ease. Their visit had given me confidence and I'd always remain grateful to them. I raised my glass and asked the gathering for a toast, from which a loud cheer "Long may he live," was heard. The toast was enthusiastically repeated.

Then the president, butcher Hubert Benz, wanted to say a few words. Hubert Benz knew Venlo really well and often commiserated with me about my predicaments. I'm sure he already heard about the opposition I had to encounter from the other butchers in Venlo once it became known I was going to establish a pork store in Venlo. "While I was meandering through Venlo last night, I asked folks here and there what was their opinion of Hubert Dielen. The feedback was such, that all previous worries I had for Hubert had faded away. And when I see Hubert's current setup and knowing the rapport Hubert has with the local people, I wouldn't be a bit surprised that within a short time, our Hubert Dielen through his own efforts will have the most prominent pork store in Venlo." A thunderous "hurray" followed president Hubert Benz's words. Then Mr. Benz asked me on the side whether I could prepare some sandwiches for the group. They had budgeted 50 German marks for such an outlay. I said, "No problem. Just have the choir stroll around Venlo for an hour so my wife and her friend Nel Seelen could get the table set."

When they returned, everything was ready. The suckling pig was at the center of the table. Two large pots of coffee were positioned nearby and plenty of rolls and raisin bread were available. Mina had everything under control and the table looked wonderful. I sliced the pig and the guests enjoyed everything to the fullest. Everyone had completed his meal an hour before the train was to depart. Hubert Benz walked up to Mina and put not 50 but 75 marks in her hand. She didn't want to accept it (the extra 25 marks) but Hubert Benz insisted. To the elation of the bystanders outside, the choir sang another farewell song. The choir now also bid farewell to Mina and then lined up four to a row in formation. As they marched in quick tempo to the train station, they were singing an upbeat march song. The onlookers enthusiastically marched alongside. At the station, the train was set to depart. I returned home after waving good-bye to the singers at the station. That afternoon was an unforgettable experience for all involved.

When I got home, Mina and Nel were eating the leftovers of the tasty suckling pig. They had worked hard that afternoon.

After the table was cleaned off, Nel Seelen went home. I asked Mina whether she wanted to stop by her father's place with me. At first she didn't want to go but then she changed her mind.

When we arrived at her father's place, Mr. Engels was at home with Piet (his brother), Jean, and Louis Mattousch, a brother-in-law of mine. They were all having a beer. Mina and I were coolly received and Louis inquired with a sardonic smirk whether we had had visitors? I sensed that that had been the topic before Mina and I arrived. I replied some friends of mine from Straelen did stop by. I got the impression they wanted to comment about it, but didn't have the nerve. Then Marie, who had just joined us, broke the silence and said, "You're really something else to have allowed all them Pruusses (an ethnic slur for all Prussians/Germans) to gorge themselves at your expense." I had to laugh at that one and replied, "Yeah, yeah, them dirty Pruusses, eh?"

Mina took offense to Marie's remark. "Yeah, they ate well and they paid for it too. They even left behind plenty for us, unlike you all after the wedding who took all the leftovers home. And if you want to know how much they compensated us, just look at this!" she said, as she displayed the 75 (German) marks that were in her pocket. Mr. Engels smiled self-consciously while chewing on his pipe. Then Mattousch remarked how the choir could've been better used for advertisement had they made the rounds at the various Venlo cafés while singing at each before having a drink. I then became angry and sarcastically replied that in the future I'd go to them to ask what I had to do. Mina, who sensed my irritation, intervened and said she wanted to go home. No one said anything anymore and we left. Mina was sorry the way things turned out. When we arrived home, we briefly discussed what had to be done the following day and then went to bed.

The following morning, as I was walking by the town hall marketplace looking whether there were any good pigs to buy, my apprentice came calling on me; He said a well-dressed gentleman wanted to speak to me back at the store. So I returned home and Mina quickly informed me the gentleman who was waiting for me in the living room was the registrar, Koenders. Mr. Koenders did introduce himself as the registrar but added, President of the German Choir Club, *Germania*. He asked whether I had a moment to talk and if I wanted to join his choir. I told him that currently wasn't possible because I needed all my energy to keep my head above water. He smiled and said how he heard a lot of praise about me from the folks back in Straelen. Feeling now more at ease, I explained I was a newcomer to Venlo and sensed a subtle hostility from the other butchers. My main problem was I didn't have enough capital. The registrar smiled and said, "Well, I can't offer you any money because I don't have any." So I floated another idea. "If I could get a few boarders to supplement my income, I'd be able to manage better." Besides, I thought to myself, I wouldn't mind having a few choir members rooming and boarding here. I may even gain a few German customers. The registrar asked, "How many rooms do you have?" "I have five rooms available," I said. "Well," said the registrar, "we'll be having rehearsal tomorrow night and I'll get back in touch with you later. I realize you can't join right now, but perhaps you'll be able to join once I get you your boarders." To that I promised. We shook hands and he left.

I then quickly went to the slaughterhouse to get a slaughtered pig for Mina to sell. Tinus Beerden also tipped me off about 2 pigs in Blerick I was able to get for a good price and immediately delivered. They were slaughtered by afternoon. Mina sold everything the following morning, which put her in a good mood.

So that previous evening I had 6 entire halves of pig hanging side by side in the store. That was something new to Venlo. In Venlo, pigs were usually cut into small pieces and then hung up. The other butchers openly snickered at my method but the public had a different opinion. It was soon known Dielen had much better cuts of pork chops and cutlets on display.

On Monday, my apprentice and I did butcher work. On Tuesday, we made sausages for Mina to sell. I also taught Mina how to present a set of cutlets in the classic way. Mina was an excellent student. I only had to teach her once before she knew how to do something. In the afternoons, I'd go to the farmers. I was soon well-known amongst the farmers. I was able to buy at more advantageous prices than the other Venlo butchers because I used a different purchasing method.

On Wednesday morning, 4 civil servants from across the border came to the store to inquire about rooms for rent. They had been sent by Mr. Koenders. Mina was taken aback by all of the sudden interest in the rooms and stepped aside. The 4 gentlemen wanted to rent by November 1st and offered to pay

32 guilders per month for room and board. Mina wasn't too crazy about having boarders in the house, not to mention they were also Prussians (Germans). So I told the gentlemen to give me some time to think it over and I'd give them an answer tomorrow. That same morning we also had some German women shopping at the store.

In the late afternoon, after Mina and I had counted the receipts of the day, Mina left to seek out new customers. Mina knew Venlo very well. She also planned on visiting her father to ask whether she could borrow some money so we could do more slaughtering. Around 4 p.m., Mina, like a true butcher's wife, enthusiastically went about town with her large cloth purse hanging from her elbow. I stayed home and took care of the store. But, oh my! At around 6 p.m., Mina returned with tears in her eyes. Mina had gone first to her father's place from which she had left very disappointed. Her father didn't give her anything because the bill from Mina's wedding had yet to be squared away. Louis Mattousch, Mina's brother-in-law, was still busy with that. Once that was settled, her father would see what he could do. For now, he didn't have anything to spare.

Mina thought how renting to those 4 Prussians wouldn't be such a bad idea after all, for then we'd have at least some regular cash flow every month. I told Mina I knew it was going come to this, so I had already told the 4 fellows they could begin renting on November 1st. Looking for new customers the remainder of the day was now out of the question. So I changed clothes and went out for a beer. First I stopped by Mr. Sanders, the cabinetmaker, and asked whether he had some reasonably priced furniture I could buy on credit and also have delivered. "No problem," he said. "Just come on over." So I told him I'd have Mina stop by the following day to pick out what was needed. At the same time, for my own business, I asked whether he'd want to place an order with Mina for a couple pounds of meat. I then sought out Jean Tendijk and Bloem Canjels with whom I negotiated similar arrangements. When I got home, I told Mina about the arrangement and she could go the following day to pick out the furniture for the 4 rooms. But she said she wouldn't go without any money, to which I answered, "That had all been taken care of."

The following afternoon, I gave Mina some names where she could get some meat orders. She didn't return until late that evening. Mina beamed as she enumerated the furniture that was decided upon. "Yeah," she said, "I now realize we can't expect much from our relatives. We'll simply have to make it on our own." So that's how we fumbled about the first few weeks of our marriage.

Mina and I had the rental rooms decorated and furnished so they were ready for the boarders by November 1st. On November 1st, we found ourselves with a house full of people. So began a new chapter in our life. Nel Seelen helped us (with chores) where she could. Mina and I, however, still had considerable financial worries. Our landlord, Mr. Wolters, expected the 1st monthly rent payment by December 1st. He felt reassured when he saw how hard we were working and even making progress.

The best room in the house was located behind the store via a hallway and was used by the boarders as their living room and they made themselves at home. The first floor above the store was rented to von den Burg, an auditor. The fellows usually played cards when they were home in the evening. They'd each buy a 6¢ mug of beer from me as needed. No wonder they enjoyed their dwelling. Even though each one had their own house key, they didn't seem to seek entertainment elsewhere. A couple of times, I encouraged my gentlemen (boarders) to join me for a beer at the Maaspoort (Engels café). But soon the Engelses were telling me how my boarders were too boisterous and allowed their cigar ashes to fall on the ground. Yeah, and one even spit on the newly scrubbed floor! Well then, that'll give me a good excuse not to bring my boarders to the Maaspoort anymore; something I can later really rub under the noses of Mr. Engels and Mattousch.

Mr. Koenders now asked me to come to a *Germania* rehearsal because they were going to have a concert in a month. I wrote him a nice letter explaining I'd allow my name to be put up for a vote. That same evening, I heard from my renters who were also choir members that I'd been unanimously voted in. The following day, I received the application forms along with the rules and regulations. I showed up at the next rehearsal. In quick military fashion, I was introduced. I knew most of the songs from years ago. Soon I was placed beside the first tenor to bolster that vocal section. The rehearsals were held in the café Beel's large hall. The café was located on Vleesstraat (Meat Street). After our rehearsal, we all drank a toast to our group. Registrar Koenders thanked me for treating the renters so well. The renters had said a lot of good things about me to president Koenders and other choir members. Now all the choir members at the café wanted to hear about my life adventure stories. I told them it was getting too late to start such stories. Whether I liked it or not, they insisted I begin. Since our choir club was registered as a society, we had a late night permit. As a result, I was there late into the night telling my stories. Mina was not too happy when I came home with our renters so late but tried to understand. She calmed down after a while.

In Krefeld, I found a new butcher assistant. I was able to prepare and preserve a lot of meats and sausages with his help. So by the end of winter, I had quite an assortment of goods on display, but I still lacked in capital. Meat purchases were at their lowest this time of the year since Lent was coming up. The town crier was ringing his bell as he walked through town yelling, "Fresh prepared pork on sale at Faldera's for 32¢/lb.!" A half-hour later was another announcement, "30¢/lb. at Haanen's place." Then a little while later, "28¢." The problem with such tactics was it cost 28¢/lb. just to buy a pig! So I decided to ask some women that Mina and I knew well to purchase for us at Faldera's and Haanen's 10–12 lbs. of pork cutlets. I was able to do this for two weeks. That cured the gentlemen (Haanen and Faldera of their ploy). The last guilder was now out of the (store's) till; bacon and ham hung from the attic to the basement.

So I continued that way for a couple more weeks but I had to have cash. Then I remembered Mrs. Kocks. She once told me if I were ever in need, she'd be more than happy to help me. So, I decided to visit Mrs. Kocks in Straelen where I found her at her castle. I was received in the friendliest manner and she was pleased to hear Mina and I were doing alright. Then I told the madam in confidence my situation, I had thousands of Deutsche marks worth of inventory but no cash. I was wondering whether she could help me by loaning 300 Deutsche marks. She said she was sorry to say the key to her cabinet wasn't currently available. Would I mind returning tomorrow? So I went back to Venlo by foot and returned to Straelen by foot the next day. The woman agreed to loan me the 300 Deutsche marks provided my father would underwrite the loan. That's the procedure her husband insisted upon. I tried to explain I'd rather not involve my father, but she didn't yield. So now I went to my father with a heavy heart.

He didn't look too happy when he saw me coming since he already knew what I was coming for. Gentleman Kocks had already related to my father my request for a loan. So I explained my predicament to my father who understood my situation and gave me the 300 marks without hesitation. I was able to take the train back to Venlo. I arrived just in time because there were only 38 guilders remaining in the till.

The wedding bill along with the trousseau arrived from Mina's house. The bill was an I.O.U. for 300 guilders, which I was supposed to sign. Apparently Mina's family felt the wedding was too expensive. I simply tossed the I.O.U. in the cabinet.

By summer I had a beautiful row of bacon and well-carved cutlets, something that was unknown here. My other meat products were also excellent, but we still lacked cash. I sat in the cellar amongst all my meats, was very discouraged and about to cry. Just then, Mina called me to come upstairs where her father and a buyer, Mr. Maseland from Maastricht, wanted to talk to me. I suddenly felt relieved. This man wanted to buy lots of bacon and other meats. Mr. Maseland already purchased all Mr. Engels had available which allowed me to provide the remainder of Mr. Maseland's needs. Mr. Engels told me how he was able to get the high price of 35¢/lb. I showed Mr. Maseland what I had available and asked what he thought. "My compliments; I've never seen a finer assortment." He asked me how much I wanted for the meats. Since I knew I needed the cash, I remarked how my stock of goods were worth 40¢/lb., but I would be willing to sell for 38¢/lb., not less. Mr. Maseland asked me to come outside and offered me 36–37¢/lb. When I didn't yield, he agreed to pay the 38¢/lb.

Mr. Maseland asked for a fine, long saveloy for his wife. He bought about 4,000 lbs. of bacon in all. Maseland opened the satchel he had hanging over his shoulder and counted pure gold coins. He looked around the store to buy some other meat for another 80 guilders. I called Mina, who was quivering with delight, to take the money. While we all had a hot cup of coffee, I mentioned that in the morning I'd get some new gunnysacks to put the merchandise in and arranged for Mr. Engels to ship everything. Just as the two gentlemen were about to leave the store, Mina suddenly flung her arms around me still quivering with joy. Mina and I had never seen so much money. You can imagine the exhilaration.

Life became much more enjoyable. The boarders were quite contented. Mina and I had no problem collecting the rent from the boarders every month and our pork store slowly but surely became more successful. When Mina was expecting her first child, Nel Seelen would help Mina wherever Nel could.

One day, I happened to be in Helenaveen (10 km west of Venlo) visiting a tobacco planter. The man told me he'd love to give his daughter an opportunity to live in a city. The girl was a bright, petite, 15 year old. Her name was Bertha van Woeziek. Bertha's mother was willing to let Bertha go to the city but first she wanted Bertha to be cleaned up and have all her clothes laundered and in good order. Perhaps Bertha had a little lice problem, which would've been unacceptable in the city. The following Monday, Bertha arrived at our store. Mina was very happy to have Bertha for a housemaid. Bertha was an excellent maid who served us for many years. She later became Mrs. van Hugten.

Mina and I were able to go to Straelen's summer kermis because we had Nel Seelen to manage the household and also a good assistant butcher. My brother Louis met us at the railroad station. I felt proud to walk through Straelen with my young wife with her flushed visage and pretty dress. She shined like a rose. In all of Straelen, I had the best looking woman. My old buddies mentioned this to me a few times and commented on my good fortune.

Then on July 6th, 1884, Mina gave birth to a strong baby girl we named Johanna. After four or five days, Mina wanted to get out of bed. I got very upset and chided her. On the 8th day, when I had to go to the slaughterhouse for a while, I returned and encountered Mina in the hallway. I was stunned and guided her directly back to bed, but it was too late. That evening, she developed a fever. As a result, she wasn't able to help in the store until three months later.

The summer had passed and after I had been in Venlo for one year, I was able to pay off all my debts and still have 4,000 Deutsche marks remaining. No wonder I was walking taller and taking the entire Engels family less and less seriously. I'd still visit the Engelses from time to time though, in addition to having a glass of beer with them and remaining on good terms with Mr. Engels. Mr. Engels reciprocated. I felt bad for Mina when I thought of how much and how hard she had worked for her family and now no longer even seems to have a parental home. Doing business with that bunch was almost impossible. They were always suspicious and never satisfied. On Wednesday or Thursday afternoon, when Mina would visit our customers who placed orders, she was always in an upbeat mood. That is until she had arrived at her family's home. Then she'd change like a leaf in autumn. Her sisters, Nel and Marie, were the main problem. They complained nonstop. They'd either condemn Mr. Engels or curse me. Even Louis Mattousch, who was responsible for supplying the officer's table (at the local military compound) and needed a lot of meat, would brush-off Mina's offers. He'd tell her the meat at Wettelings' place was better and cheaper. Then Mina would return home in the evening to tell me, as she cried, about the inconsiderate treatment she had received from her family. I told Mina more than once not to go back there anymore, but that didn't help. The following week, it was the same old thing again. Finally, I forbid her to go to Mattousch and that worked. Mina now avoided socializing with her family as much as possible. As a result, a lot of the vexations were removed.

My father died in Straelen November 23, 1884. I did my best to handle the family's estate in a peaceful manner. Louis continued to reside in the house and kept ailing Philip with him. We didn't have an estate sale. Drika and Philip selected some items they preferred, and I chose something for the sake of my own memories. The remaining items, which included the house and garden, were then evaluated and Louis became the owner. Another parcel of land, which was about 8 acres, was publicly sold. After everything was squared away, I caught the last train home.

The business was doing very well. One day I heard that my neighbor, Mr. Haanen, had serious financial problems and as a result, would be putting his business/house up for sale at the local auction hall. I offered him 7,000 guilders but Mr. Haanen declined. Two weeks later, he changed his mind and said he'd accept my offer. But after thinking it over, I had lost interest in the proposition. For the last three years, I'd been depositing the profits of my business in my account at Mr. Berger's bank and as a result, had accumulated a nice sum. Through Mr. Berger, the banker, I heard it wouldn't be long before Mr. Haanen's business/house would be sold under article 1223 of the civil law code (Chapter 11, in today's U.S.A.). When that occurred, I should be ready to make my offer if

I was still interested. I'd be able to get the business/house for a much lower price. And sure enough, after a couple of weeks, the business/house was to be auctioned off by court order. So I sent my carpenter, Mr. A. Deltrap to the auction with firm instructions. I stayed at home and waited. Everything went smoothly and Deltrap was able to get the house for 5,100 guilders!

After the paperwork was done and I became the official owner, I had the house completely renovated to my own specifications. I transferred my business/home to my own building of course, to the chagrin of Mr. Wolters. I didn't have to move but Mr. Wolters didn't want to make the needed wear and tear repairs. That's the reason we decided to move, even though our new residence would be much smaller.

Louis loaned me 1500 Deutsche marks and due to that, we didn't have any financial worries. My business continued to grow and after a couple of years, one could say I had just about the best pork store/butchery business in Venlo. Many customers came from Germany. The custom officials at the border allowed travelers to take 4 lbs. of meat back across the border to Germany.

Just when everything was going so well, I was dealt a setback I barely overcame. On January 10th, 1890, Mina died. She was only 37 years old. It happened after the birth of our fifth child. I was confused and totally devastated. When Mina was carried out of the house, I almost collapsed. My brother Louis held me up during the funeral, but after the ceremony I withdrew and didn't want to see anyone. I was simply distraught and didn't know how to continue on. What was I going to do now? I had 5 small children, 4 assistant butchers, 2 female sales clerks, and a cleaning woman. Not to mention that we also had a very busy business in our new, rather cramped, quarters. A Mrs. Houben of the Veen (tr., slough) was willing to take into her house my two youngest children. As a consequence of all the disorder, I experienced more hassle than usual with my workers who were mostly Germans of the worst sort. One time I had even hired a servant who, unbeknownst to me at the time, was a thief. It was a good thing I was of strong built and not easily intimidated.

Nel Seelen did everything she could to assist me. But all the help still didn't prevent me from being swindled and robbed. Later on, I discovered the biggest thief was the washerwoman, Hermans. I caught her at 1 a.m. in the store and had the police take her away. A search of her home yielded a wagonload of hams, sausages and linens. My courage and willpower left me. The business had grown to such an extent that we'd slaughter about 25–30 pigs a week. But by now, we were only able to handle 15 pigs a week and still had a hard time managing everything. I felt as if I was getting poorer by the week. Now and then, I'd visit Mr. Engels with my children Gerard and Louis, but such visits didn't give me any consolation so I'd simply return home. Then on Saturday, Mina's sister Marie would come and help a little. But she'd recommend to the customers who lived closer to the Engelses to shop at the Engelses' store. That didn't do my business any good, so I told her to stay home.

Then Hubertina Engels offered me some help. She was just the opposite of Marie. She was calm and good-natured. That worked out much better. When I paid another visit to Mr. Engels on Saturday night, Hubertina said, "Father, things aren't going well at Hubert's place. Hubert must find help." But Mr. Engels didn't have any answers.

VII. Hubertina (2nd wife & sister of Mina)

Every morning, Hubertina would stop by to help. She even offered to take in my two boys (to be cared for at the Engelses' place), but that I'd never grant regardless of my situation.

Time continued to march on and I became poorer by the month. One evening, I was at Mr. Engelses' place again feeling very depressed. Mr. Engels did his best to lift my spirits but I knew I'd go bankrupt if something didn't change. Hubertina said once again, "Let me take care of the two boys. My fiancé died and I'll raise the boys as if they were my own." At first I said, "No way." But then after thinking it over, I said, "Listen, Hubertina. You can have the two boys, but you'll have to take me as well. I'll give you 8 days to think it over. You all know me well enough by now. I'll be back Saturday night to hear your reply; and if you decline my offer, I'll simply go on a trip Sunday and not return until I've found a wife." Mr. Engels sat stunned. Hubertina quickly looked around for a chair. No one else was there. "So, until Saturday; good night." What occurred later that night at the Engelses' place once I had left, I don't know. The days following this conversation, Hubertina didn't come to the store. I thought my offer may have fallen through, but I simply couldn't continue the way I was going.

The week proceeded along and then Friday evening, stood Hubertina at the door. She said, "I'll come tomorrow," then immediately left. I must say that for the past several weeks during the weekends, I had been going to the nearby Catholic shrine of Kevelaer to fervently pray for assistance. That seemed to have worked!

The following morning, Saturday, Hubertina arrived in a happy mood at her usual time and began her routine, which didn't end until evening. I escorted her home as I always did and on our way, Hubertina said, "Hubert, I have decided to stay with you from now on. You get my hand and heart. Just come to my house to get my father's approval as well." So that's what we did.

When I entered the room, everyone (except Mr. Engels and Hubertina) got up and left. Then I said to Mr. Engels and Hubertina, "I've come to get your answer." Mr. Engels said he inquired with the monsignor about us and the monsignor had no objection. So if it were all right with Hubertina, it'd be fine with him too.

Mr. Engels wanted me to be aware of the Engelses' current dilemma were Hubertina to assert her claim on the Engelses' home. Why? Because the title of the Engelses' home was under the name of Tinus, Hubertina, and Marie; the remaining Engelses would then have to liquidate everything just to pay off Hubertina and as a result would end up on the street. So I said, "Mr. Engels, your current arrangement can stay as it is. I'm only interested in finding a good wife and I believe I've found one in Hubertina."

I want the reader to be aware that two years ago, Mr. Engels had vouched for Jean in a loan. Jean Engels ended up going bankrupt. Mr. Engels made an effort to protect his three other children from his own possible bankruptcy by transferring the title of the family's home to the names of Tinus, Hubertina, and Marie. Nonetheless, Jean couldn't resolve his problem and ended up in debtor's prison. Mr. Engels was now obligated to discharge Jean's debt. Mr. Engels claimed in court that the names of Tinus, Hubertina, and Marie on the title of his home couldn't be reversed. I don't think such maneuvering was even legal.

Hubertina said, "I hope I don't regret this big step I'm now taking. I'm doing it mainly out of compassion for the children." We shook hands and the others now came out of the kitchen to congratulate us. Everyone toasted one another with a glass of wine and then I went home. Everything was quiet on Sunday but by Monday morning, Hubertina was back at work.

After a couple weeks passed, Hubertina and I notified the city registrar of our intended marriage date. So there I was, again applying for a marriage license from the religious and secular authorities. After Hubertina and I had done that, I checked up on

Anna Marie Hubertina Engels 11/1/1857-8/7/1941
1915 photograph

the town hall a few days later to make sure everything was going all right. Sure enough! The marriage license from The Hague hadn't arrived. At my insistence, mayor Houben immediately sent a telegram to the ministry of interior affairs. The mayor and I received a reply at 4 p.m. which stated, "Marriage can proceed; certificate/license will follow."

The marriage certificate arrived by express mail the following morning at 10 a.m., on July 28, 1890, at the town hall and was placed on a table. Hubertina and I arrived at the town hall at 10:30 a.m. for our civil marriage but the mayor wasn't there. The alderman, Peters, said he'd perform the civil function since he was the acting mayor while mayor Houben was out of town. As I handed Peters the marriage certificate/license, he said, "Now that all the necessary documents are present, mayor Houben could just as well have performed the function himself. Mayor Houben had no need to evade this marriage by going to Blerick." Peters explained how the mayor was only allowed to marry couples who had all of their documents together. Otherwise, the mayor could be held liable. So now if something went awry, Peters would be the fall guy. Peters, Hubertina and I entered the city council-chamber where a quick civil marriage for Hubertina and me occurred. Hubertina and I then proceeded to our church ceremony. There, 3 priests celebrated High Mass and thus was our marriage blessed. Hubertina and I tried to keep the whole wedding ceremony as low-key as possible; hence, there wasn't much revelry. Hubertina and I departed at 7 p.m. for Koblenz where we enjoyed a beautiful 3–day honeymoon cruise on the Rhine.

When Hubertina and I returned home, we went straight to work. We soon got everything back on track and moving. My two youngest kids were retrieved from their caregiver since I now had a pretty good maid, an excellent nanny, and also Nel Seelen frequently was at our place. It wasn't long before my kids were soon addressing Hubertina as "Mom."

Hubertina had a completely different disposition than the late Mina. Hubertina's philosophy was that if she was nice to someone, why would they be mean to her in return? That's all very noble and everything but a little naïve.

After a couple weeks, Hubertina, like Mina before her and temporarily Nel Seelen, would seek out our usual customers and any others for the possible placing of orders. On Thursday afternoons, Hubertina would return quite happy and satisfied with her results. She had often gained additional customers. But on Fridays after returning from her own family and ultimately Mattousch, she'd be in tears. That's when I put a decisive halt to Hubertina seeking to place meat orders from amongst her family members or Mattousch. She was allowed to make private social visits but nothing pertaining to soliciting meat orders. Fortunately, Hubertina and I could survive without their business.

Hubertina and I had problems with the two youngest children. They didn't fare well at all and within one year, they both died. They were now our sweet little angels in heaven.

Soon thereafter, on February 17, 1891, Hubertina gave birth to a baby boy we named Hubert.

In Straelen, all was going well. Louis had finally gotten married. His wife's name was Stiena Theunissen. She was a sweet and gentle woman who raised her children in complete piety. Some time later, our neighbor in Straelen put his house up for sale. It was a brewery named Briemens.

Louis now asked me to settle my remaining 1,000 Deutsche mark debt. This created a lot of bitterness between us. I must say I wasn't all that fair to him. The origins of this conflict were as follows; not long after my mother had passed away, my father's sister, Mieke Tante (Aunt Mary), came to visit the Dielen household in Straelen. This event occurred just before I moved to Venlo. After a couple of days, my father wanted Mieke Tante to live with us in our family home indefinitely. Louis, Drika, Philip and I were going to make sure that didn't happen. Mieke Tante was a good Christian woman but having been an old housekeeper for a pastor, she was opinionated and unbearable. She continuously complained about her late husband and how he had spent most of her savings. At first, Father wouldn't hear about sending Mieke Tante back to Kessel, Limburg since he'd been the caretaker of what little she'd been able to stash away, an amount he couldn't remember, and consequently felt obligated toward her. Louis, Drika, Philip, and I came to an agreement that we'd make sure she wouldn't lack for anything. We then put Mieke Tante with ample provisions in the ponycart and sent her back to Kessel.

As soon as I had settled in Venlo, I arranged with Drika and Louis (Philip having recently died) that I'd send a package of

The four oldest children circa 1892

victuals to Mieke Tante once every two weeks. I'd then present Louis once a year with the bill plus interest and he would in turn deduct that amount from the debt I had with him. Drika, however, didn't live up to her end of the bargain, even though I had reminded her at least ten times.

One early morning, a ponycart stopped in front of our door and there on top of a pile of straw in the cart sat Mieke Tante with her large gray bonnet. We helped her out of the cart and the cart went on its way. Mieke Tante took off her cloak and began assisting us in the store, at least that's the impression she wanted to give. Hubertina's face became more sour by the minute. That night we found Mieke Tante a bed and the following morning she began giving orders to everyone, especially to my wife. In addition to that, Mieke Tante thought she had to criticize everything she thought was done wrong.

After the noontime dinner, I told Mieke Tante to get her coat. I told her I had to go to Reuver by train and I thought it best if she'd come along so she could return to Kessel. That didn't go over too well and as Mieke Tante cried, she used every sort of pretense in order to stay. In the end, she even maintained that the bread in Kessel was so bad she couldn't digest it. Even the coffee beans in Kessel were no good. I promised her that every week I'd send her fresh milk and coffee beans from Venlo. Her farmer basket was filled with food and a half-hour before the train departed, the housekeeper and I escorted Mieke Tante to the train station. We were all very sad; now I'm lying. Once Mieke Tante and I arrived in Reuver, I was fortunate to find someone with a cart who was on his way to Kessel and was willing to take Mieke Tante along.

When I went to Straelen, I told Louis about Mieke Tante and what occurred. Louis bursted out laughing and said, "Don't feel bad about it, Hubert." But I reminded him that he practically inherited the entire family estate, which also included certain obligations (such as assisting Mieke Tante). Louis chuckled and replied, "That's what you think." Then Louis mentioned I still hadn't paid off my 1,000–mark debt. At the time, I was reluctant to resolve the debt because I was strapped for cash. I remembered what Hubertina advised, in that if we settled the 1,000–mark debt, we would probably end up being the only ones left sustaining Mieke Tante. I thought that plausible at the time, but in hindsight

I should've settled the debt just to see if Louis would've reimbursed me for half of Mieke Tante's expenses.

One day, Mr. Theunissen, Louis's father-in-law, stopped by my store. Mr. Theunissen and I had always been good friends and so in that spirit, I frankly aired the contentions that existed between Louis and me. After I told Mr. Theunissen the whole story, he said, "Yes, Hubert, your story is quite different from what Louis said." I once again insisted this was the truth and asked Mr. Theunissen to mediate this conflict between Louis and me. I wanted a peaceful settlement. Mr. Theunissen doubted he could change Louis's viewpoint because Louis was quite stubborn and still resentful at me from his last visit to us. He was then staying with us because of his lawsuit with his neighbor Muisers. Louis asked my opinion and I suggested he put himself in Muisers' position. Louis would then realize he'd lose the case, which is what happened nonetheless.

After a couple of days, I received a letter from Mr. Geelen, a lawyer. Mr. Geelen asked me to attempt a resolution with Louis before the situation got out-of-hand. I prepared a bill of all the expenses incurred by Mieke Tante, deducted that amount from my 1,000–mark debt and let the Berger Bank mail the remainder to Louis.

To the lawyer I related how things were proceeding and that I sent the money (to Louis). To Louis, I mailed a letter with a statement of the account and that the remaining balance would be transferred by the bank. I also wrote of the pleasant visit from Louis's father-in-law and my reciprocal visit to Mr. Theunissen seeking his assistance in mediation, which unfortunately didn't succeed.

Now for a long time I didn't hear anything from my family (Louis and Drika) until the death notice of Mieke Tante came. She died January 19th, 1900 at the age of 77. I had faithfully sent Mieke Tante her weekly package right up to her death. I immediately notified Louis of Mieke Tante's death and was going to leave the following morning at 10 a.m. for the hamlet named Leuman, located near Reuver, where Mieke Tante had resided.

The train coming from Straelen connected with my train going to Reuver. Louis came out of the train (from Straelen) and we politely greeted each other. I asked him whether he was on his way to Reuver and he nodded. In Reuver, Louis asked me where

Mieke Tante had lived. He had never been there. I knew where she had lived so in no time we were at her house of mourning.

One of the neighbors, who had recognized me from a previous time, came outside her house and immediately began lamenting about all of the distress she had encountered with that old woman. She led us toward the house and we entered through the kitchen toward the room where Mieke Tante was lying in state. Mieke Tante looked as if she was still alive. Another neighborhood woman came in chiming with the first about how demanding Mieke Tante had been. I asked the two women whether such a topic was appropriate at the moment. The two women got the hint and talked of something else. Apparently, they'd been present at Mieke Tante's final moments and mentioned Mieke Tante's request that the small blue linen satchel be given to Hubert Dielen. Another neighbor, Mr. Gielen, had Mieke Tante's bankbook. As the women were called for their noontime dinner, they asked whether Louis and I wanted to join them. Louis and I politely declined but offered to join them an hour later for a cup of coffee.

Once the neighbors had left, Louis and I gave everything a quick look over. There was a cupboard, which held pots and pans; a table with three rickety chairs; and a bedstead with pallet, feather-filled-cover and some sheets. In a room, there were a few more chairs and a wardrobe. Amongst the wardrobe, I found the last package I had mailed to Mieke Tante. "Well," said Louis, "If she had received such a package weekly, then surely the neighbors had partaken in it." Besides the few aforementioned items, there wasn't anything worth keeping. I asked Louis what we should do with all this junk. Louis said, "I don't want any of this." In the small satchel was still a total of 60 guilders and at her neighbor Gielen's, who had the bank booklet, I counted 275 guilders.

In the meantime, chaplain Bloemen stopped by. I told Louis that the chaplain sort of watched out for Mieke Tante. That was something Louis didn't know. The chaplain remarked on how considerate it was of me to mail Mieke Tante such a package every week. The chaplain had also spent 4 or 5 guilders on Mieke Tante. I compensated the chaplain 5 guilders from the satchel. As far as the funeral, I had that arranged with the pastor. The chaplain then left. Louis conceded he didn't realize how much I had done for Mieke Tante's welfare. I shrugged it off and said I sort of considered it my duty to care for a relative. I proposed that since Mieke Tante was such a burden to her neighbors, we'd compensate them by giving them the furniture. The contents of the satchel would be used for the funeral and Holy Masses. Then we'd split amongst the two of us the sum in the bank booklet. We could forget about Drika, who never showed much concern for Mieke Tante anyways. Louis said tersely, "That's fine with me."

Louis never was a man of many words and now was no exception. Such laconicism was getting on my nerves and so I said either he let me know what he thought of this current matter or I'd return to Venlo. My remark made somewhat of a difference in that Louis then complimented me on how I had managed everything superbly and wanted me to continue. I said, "That sounds better, and if it's all right with you, I'll take care of the funeral arrangements too."

Louis and I walked back to the house of mourning where the two neighbors and the landlord, who was also a neighbor, were waiting near the coffee table. Louis and I told them that out of gratefulness for their assistance to Mieke Tante, we wanted to give all of Mieke Tante's belongings to them. Our one condition was the belongings be apportioned without any arguments and it be done as soon as Mieke Tante's funeral was over. They all gladly agreed to that. I asked the landlord, who was a carpenter, how much a coffin would cost. He said 7½ guilders, plus 1 guilder for additional expenses. I gave him 10 guilders and asked him whether there were any other outstanding bills. Yes, some back rent of 4½ guilders but he wanted to let it go. I refused and offered him 5 guilders. "Is everything now taken care of?" "Yeah, sure," he replied.

Now it was coffee table time. The women had prepared bacon, eggs, and a good pot of coffee; all of which Louis and I had to partake whether we liked it or not. "Mieke Tante was quite demanding, but you (by giving the belongings) have made up for it," they said. After the coffee, we thanked the little women and said, "Till the day after tomorrow." Louis and I now went to the town hall (for notification of Mieke Tante's death) and then to the pastor. We introduced ourselves to the pastor and were formally received. The pastor thought the money in the blue satchel should be sufficient for a decent funeral and 9 Holy Masses. I gave the satchel to the pastor and said, "Mieke Tante has certainly provided for her own funeral expenses." A good bottle of chilled wine was at hand and as we enjoyed the contents, the pastor said the little satchel was proof someone had taken good care of Mieke Tante otherwise she wouldn't have had so much left over.

Two days later, Louis and I returned to Reuver. We didn't say much to each other (on this short trip). Even though the train was a little late, the pastor waited for us. The funeral Mass for Mieke Tante was officiated by three priests. After the funeral, Louis and I invited everyone at the funeral for a drink (at a local restaurant), since that was the custom. I thanked everyone for showing their last respects to the deceased. Then Louis and I stopped by the post office to retrieve Mieke Tante's remaining savings. We then went to the train station where we barely caught the last train to Venlo.

Once we arrived in Venlo, Louis had to wait for more than an hour for his connecting train to Straelen. I gave him his share of the Mieke Tante savings and invited him to come to my house. He declined. So we said good-bye and I left for home.

Two months later, Drika stood in front of me. We conversed about Mieke Tante, of course, and Drika's share of the inheritance. I said to her, "You too are from Straelen; hasn't Louis told you anything?" After initially playing dumb, Drika then acknowledged she'd been somewhat informed about the dispensing of Mieke Tante's legacy but thought it unfair. I pulled out my folder of letter copies and read to her a couple of the letters I had written her. Startled, Drika said she'd never seen any of those letters, her husband must have withheld them. I then asked Drika what had been her opinion of the late discord between Louis and me. Drika felt I was in the right. Drika now told me about how her husband drank too much and made her life miserable. I told her I already knew that and that was the reason I never went to Borbeck. Before Drika left, we stuffed her suitcase full (of goodies) and also gave her a package of choice meats. Drika left and a year later she died. Louis asked whether I was going to go to the funeral with him, but I refused for fear of what I might do to the man who had so mistreated my sister. I instead asked Louis to give Drika's husband, Mr. Hendricks, a letter I had especially prepared. Something of which I later heard Louis actually did.

The years went by and I would visit Straelen less and less because as soon as I saw my last living brother, I was instantly reminded of our on going enmity, which I found to be a heart wrenching experience. As a consequence, I discontinued going. Then one day, Louis's wife, Stina, stopped by. Stina and Hubertina easily got along. She asked Hubertina and me to visit Louis and her the upcoming Sunday morning. Hubertina and I naturally accepted but arrived in the afternoon of the agreed upon Sunday. We were kindly received. After all these years, Louis and I were finally able to carry on a decent conversation. Louis and I were glad the enmity between us was over and we were back to our old friendship. We never again brought up the disputed subject.

During this time in Venlo, something new occurred. Since Hubertina was now with me, the Maaspoort had hired a capable replacement. But then what happened? Tinus had fallen in love with the girl. When Marie became aware of this, she did everything she could to thwart the ongoing rapport. Marie even went so far as to disallow Tinus's handling of the cashier, even though Tinus was considered the boss in the house. So Tinus stopped by our house to borrow a few marks so he could buy a glass of beer. One evening he was even locked out of his own house. The following morning, Tinus stopped by our house with tears in his eyes to tell us about his situation and asked me to come over that evening. Later that same morning, Marie also came to our store to tell us there was a problem at the Maaspoort and would I be so kind to come over in the evening.

That evening, Mr. Engels, Marie and Tinus were seated in the living room awaiting my arrival. I asked why they called me. Mr. Engels began by mentioning all the questionable things Tinus had lately been up to, saying how Tinus now even wanted to marry the new help. Marie then inveighed against Tinus and the new girl. Marie formerly had nothing but praise for the new worker. Tinus, who happened to be in a stupor, sat whimpering in the corner and only managed to say a few words.

Then I spoke, "You all have summoned me for my opinion, so now I'll tell you what I think. Tinus could easily find another girl, but at his age that is for him to decide. After all, one doesn't get married for someone else, but for one's own sake. Up till now, I've heard only how reliable, industrious, and pleasant the girl was and attractive as well. What Mr. Engels has stated is somewhat true but he should remember that since he's no longer the owner, he can't call the shots anymore. As far as Marie is concerned, if Marie had locked me out of the house; I'd have kicked in the door, grabbed her by the neck, and thrown *her* out of the house for a change. Then she'd know who the boss was. If I were Tinus, I'd even do it now!"

Nobody said anything. Tinus sobbed like a small child. "You all now know what I think and so I'm going home." By the next day, the word had already gotten around of my visit to the Maaspoort and everyone gave his or her opinion for or against. A few days later, I heard that Tinus began drinking a lot. Actually, he was constantly drunk. I feared the worst for him. He wasn't very stout. He was a good guy but had a weak constitution.

On a later day, I was again called upon to visit the Maaspoort in the evening. Tinus wasn't well. As I entered, I saw the dejected facial expressions and knew something was really amiss. Arriving upstairs, I saw Tinus lying on his bed and he offered me his hand. He seemed to have delirium tremens. He saw monsters wandering everywhere; in bed, in the room, through me and toward the wall. He spoke incoherently. He did say that in his bed was a satchel with 1,700 guilders and whether I'd take it along. I dismissed that idea.

The next day, the friars asked the Engels family members to participate in the prayers for the dying. Someone had apparently prearranged with the notary to have Tinus write a will at that time. On December 11, 1895, Tinus died at the age of 40. That evening, Piet Engels, who was Mr. Engels's brother, invited Mattousch and me to go with him to the Maaspoort. Sitting there, Piet began by asking us whether Tinus had left a will. Mr. Engels interjected, "Yes, and Marie is the only heir. So you people no longer need bother." Piet was completely flabbergasted on hearing that. Mattousch also couldn't believe what he was hearing and was about ready to leave. I sat quietly in a corner and wanted to first drink three beers, which were reluctantly provided.

I then brought up the former verbal exchange I had had with Mr. Engels, Tinus, and Marie. Piet and Mattousch were hearing all this for the first time, which resulted in their inveighing against Mr. Engels. One could see that Mr. Engels had had enough of the quarrel as he sprung up and went to the front door where he beckoned the police to come in. The police had shown up to keep back the curious crowd the quarrel had attracted. The police requested Piet, Mattousch, and I leave but I suggested the police leave. The police then returned with their police chief. Mr. Engels told the police to remove Piet, Mattousch, and me from the house and to write up an official report citing domestic disturbance.

I then told the police chief that Mr. Engels was no longer the owner and was simply the recipient of his family's kindness. I said if anyone had the most weight in the matter, it was I since I married two of the daughters in the family. The police chief, who didn't seem to care much for Mr. Engels, then understood the confusion and departed with his constables. Following the police involvement, I was even more aggravated. When we all reentered the house, we shut the doors and windows so the public could no longer hear. I again inveighed against Marie pertaining to the former evening. I just couldn't stand that outspoken, repulsive, sister-in-law and when I recollected how she had harassed the good-natured Tinus, I could no longer contain my anger. By the time my outburst finally came to an end, Mr. Engels and Marie were sobbing with their heads bowed on the table. Piet, Mattousch and I departed.

I must say I do regret any of this had ever happened. The onlookers took great pleasure in Mr. Engels's plight since he had yet to attain much goodwill in the neighborhood. (When a death occurred, funerals were usually planned 3 days later in the late morning, followed by a brunch for the immediate family and close friends.) After the funeral, everyone went home right away (rather than the usual follow-up brunch).

A little while later, Marie got married to a butter/(creamery) merchant named A. Ploem. Hubertina then went by herself to the Maaspoort to inquire about her inheritance. But Hubertina was so rudely received, she immediately turned around and left. Although when I married Hubertina I said I only needed a wife not money, it was understood the money would be settled later. Then one day, I was asked to visit the lawyer Mr. Wijnand, where Mr. Ploem had consigned a thousand guilders. Hubertina and I thus received 975 guilders.

In February 1897, Mr. Engels had a fatal heart attack as he ate his noontime dinner.

The youngest Engels daughter, Gon Engels, was married to a Jos Zweiszpfennig. They had had a business in Eindhoven. Their business performed splendidly in the beginning. That was way back in 1886. After all, Eindhoven was then an up-and-coming location, which had yet to have its own competent pork butchers.

I would've loved to have begun my business in Eindhoven had only Mina been willing to transfer there. Jos and Gon had obviously heeded my advice when they decided to begin in Eindhoven. But Jos was a big talker and went out a lot, where as Gon was insolent and ill-mannered. Consequently, they went bankrupt and quitted Eindhoven in failure.

Hubertina and I no longer heard from her sisters and I didn't want Hubertina to ever bring up the subject of her family again. Everything with our immediate family was fine. We worked hard and made good money. Whenever I was in town at noontime, I would stop by here and there for a shot of gin. But I would usually stop by a café that was owned by one of my customers and always ended my excursions by stopping at Piet Seelen's place on the Parade (a plaza). Since I was quite a raconteur, that drinking-hour was well attended and there was a lot of laughter. But the downside was we would drink more and more until eventually the drinking-hour, that originally lasted until 1 p.m., would stretch into 1:30 p.m., and then even to 2 p.m. And so consequently, when I'd arrive home, I had had quite a few drinks in my system.

It was fine with my wife that I went out, but she also wanted me home in time. I tried to stop after two or three shots of gin so I'd be back at the latest by 1:00 or 1:30 p.m., but that didn't work out. During the drinking hour, we had such a good time our jowls ached from laughing. One day I came home and Mom (that's Hubertina) said, "I'll bet tomorrow you show up at 2:30 p.m." I didn't answer but her remark had an affect on my appetite. Sure enough, the following day I arrived at 2:30 p.m. Hubertina brought me my soup without saying anything. Everyone else had already finished their meal and returned to work. Then Hubertina said, "Wise up, man! It's already 2:30 p.m. I don't begrudge you your drinking hour, but at least arrive in time for the midday dinner." I was angry with myself and replied, "Shut up, or I'll throw this bowl of soup in your face!" My wife, of whom I'd never before insulted, began crying. I apologized and lucky for me, my apology was accepted. I had lost control of myself and subsequently swore off all hard liquor. The following day at 11:30 a.m., my wife called out to ask if I was not going to have any nips of gin. I said to her I was going to change my clothes and walk for an hour but I have had my last shot of gin (yesterday).

I didn't allow any of my workers to smoke in the sausage-workshop, including me. So I'd smoke my pipe as I walked out of town toward Genooi (a religious pilgrim village), which is by the Maas River (on the way to Grubbenvorst). After saying my little prayer in the chapel there, I stepped outside, lit my pipe and strolled calmly back home. As I went through the Villapark and came upon the Hogeweg (a country road), whom do I see? There in a nearby garden plot was my wife's cousin, Gerard Ham, gardening. As I walked onto the garden plot, I knew right then and there what I wanted.

VIII. Venlo

I wanted to work in a garden like that too, since it'd be an excellent outlet for me. I wanted to buy such a plot of land at any price. So I asked Gerard from whom did he purchase the land? He said he was renting it for 30 marks per year from Mr. van Aken. With this newfound information, I went home in a happy mood. That evening, I strode on over to Mr. van Aken. I had 200 guilders and a bill of receipt tucked into my coat-pocket since I was determined to buy that garden. I returned home with my now signed bill of receipt stating my 200 guilder down payment toward my 1300 guilder purchase. Mom (that's Hubertina) was very happy, since I'll now be working in my garden rather than drinking with my buddies. I refunded Gerard Ham his 30 marks and then I began. My transition to gardening (from my noontime buddies) wasn't as easy as I thought it was going to be, but I got used to it.

I built on my new plot of land a small cabin for my garden tools, tobacco jar and tobacco pipe. Then I drew on paper a diagram of how I wanted my garden to be. Once I was satisfied with that, I began working. It being autumn, I couldn't do much damage. First I dug a shallow trench for the garden pathways that conformed to my outline. Then I had a fellow from Venlo bring in some gravel of which he placed onto the pathways. After that I planted some small palm trees alongside the paths. When early spring came around, I had the Lenders nursery from Steyl (near Tegelen) deliver an assortment of long-stemmed roses, some coniferous trees and fruit trees. And so by the time spring arrived, I had the most beautiful little garden in Venlo. It's a pity my wife didn't share my interest in the garden but she was entitled to her opinion. She'd just say that her place was in the home.

Our children were healthy and every year the stork would bring another child. When Johanna was 8 years old, I brought her to the boarding school (of the Ursuline Sisters) in Grubbenvorst where she stayed for 6 years. Gerard and Louis attended for 6 years St. Hadelin College in Visé, Belgium. So thereafter, all my daughters would attend school in Grubbenvorst and my other sons would attend school in Visé, with the exception of the youngest, Jean, who due to the outbreak of The World War (WWI) attended the Diocesan College in Weert and then went to Rolduc (a seminary near Maastricht). Consequently, all of my 10 children, 3 from Mina and 7 from Hubertina, ended up with 6–8 years of (private) schooling. That cost me quite a bundle.

When Johanna concluded her boarding schooling, I sent her for further training to an old friend of mine, Bernard Rottman in Krefeld. He probably had the best business on the Rhine. Gerard, Louis, Hubert and Willy also served their apprenticeship there.

My monthly visits to my customers in Eindhoven, Helmond and Deurne became increasingly difficult as soon as I had discontinued drinking because it was expected during the bill collecting that I was to join them in their downing of quite a few shots of gin. Hubertina offered to free me from that problem. After all, she was the bookkeeper. So we gave that a try and although she had to endure some snickering in the beginning, she did very well. There were times when she had to collect 1400 guilders and even returned with 1300 guilders; whereas I'd be lucky if I collected even half of that, not to mention the money spent on drinks. So it all turned out for the best.

On the market plaza was a narrow house up for sale, which used to be *The Black Horse* café. I could've bought that house for 1300 guilders with the goal of later purchasing Paulussen's house, which was beside it on the corner. Then I could've established an inviting butcher business there.

My garden was still my favorite place. During the rain or other bad weather, I'd often sit for hours in my small cabin as I contemplated my estate. It wasn't long before I had inexpensively acquired five adjacent plots of land. But just like their mother, my children didn't take the slightest interest in my increasingly beautiful garden. No one was eager to assist me. In fact, I even encountered hostility and outright verbal abuse.

At this time, mayor van Rijn mentioned to me the possibility of the city expropriating a strip of my garden in order to build a street. I acceded to that prospect provided the street was named *Dielenstraat (Dielen Street).* My garden consisted of 34 plots of land. There was another garden parcel I wanted to buy, owned by a Mrs. Raub, but it was overpriced. As a result, my plans unraveled. So I decided to sell my entire 34 parcels en masse. But who'd want to buy such a large property? How does one go about finding such a buyer? That was going to be a challenge!

But one afternoon, luck was on my side. Because as I sat in my little garden cabin smoking my pipe and drinking coffee, the parish dean and an architect named Seelen stopped by. After they looked over my garden, it became evident the parish dean wanted to purchase the entire garden. The dean was going to build a gymnasium (college preparatory school). Seelen, the architect, had his doubts, which obliged me to use all of my persuasive faculties. I thus induced the dean into buying the whole garden, clearing a profit of 8,000 guilders. So that was squared away. My (later and) last garden plot purchase was from an acquaintance named Jean van Willijk. He sold me the little garden, which was about 3/4 of an acre at the Helbeek (in Venlo), for 1250 guilders. That was large enough for me.

My business at home was thriving but the house was getting too small for a growing family. The gentlemen boarders, who transferred with us from the previous house, had moved on. And now my older children, who just completed their boarding schooling, were returning home to partake in the family business. Consequently the bedroom accommodations became insufficient. An adjacent house became available on two previous occasions, but in each instance I was short on money. As a result, Wolters, who lived on the other side of the available house, bought it.

I had earlier asked Paulussen whether he was interested in selling his corner lot, which was across the street from me, but he said he wasn't ready to sell yet. (That corner was later bought by Bervoets, who turned it into a large clothing store.). Sometime later, the old couple Wolters had died and the children began the apportioning of their inheritance, which included house number 14. I heard the Wolters heirs had no intention of selling their real estate and were simply apportioning ownership titles within the family. So I made a good offer nonetheless (on house #14, the house adjacent to mine) and hoped for the best. After all, since I had 10 children to take care of, perhaps God could now assist with housing.

One morning I stood at my front door in a pensive mood when Emile Wolters walked by. He said, "Dielen, you should buy our house." To which I replied, "Sir, you're asking too much for it." He then said, "Today my brothers and sisters are getting together to allocate everything and it'll all be settled before the end of the day. So if you're still interested in buying the house, say it now. We've always been good neighbors and I'll defend your bid if you want. After all, I'd rather see you get the house than some stranger." I repeated I wouldn't mind buying their house, but they were asking too much. A fair price, I'll pay. He asked what I thought would be a fair price. I said, "12,000 guilders." He laughed and said, "For that price, I'd buy it myself!" Then as Emile walked on, he said, "I'll have someone call you when we get to that subject. In the meantime, mull it over."

I told Mom (Hubertina) the Wolterses were considering selling the adjacent property to us, which elated Hubertina. I didn't have my usual midday appetite (due to the excitement), so I rested a little as I continuously reviewed in my mind what I had to do. By 3 p.m., I was sitting in the Wolterses drawing room. Emile asked if I'd been thinking about the house and I said he knew well I had but I didn't know what the Wolters family thought. Emile said his family deemed the house to be worth about 15,000 guilders. I told Emile I don't have that much money. I could offer a little more (than 12,000 guilders), but not much. Emile said, "Give me a moment; I'll consult my family once more." In the beginning I was disheartened, but I now sensed a slight chance. Emile called me over to the family room where the Wolters family was sitting with the notary Mr. Muller. I greeted everyone and then we were each allowed to give our respective opinions.

Finally I spoke. "Here I stand, a poor butcher with 10 children, facing a group of rich people. The reason I want to buy the house is because I don't have enough bedrooms. Just to save a few guilders, I have to toil day and night for everyone (in my family). My neighbor Emile can attest to that." And thus, I continued

Hubert Dielen's 50th birthday in 1905 with his entire family. (The first row from the left is Regina, Petronella, Jean, Hubertina, and Marieka. Behind Marieka is Hubert, and then to the left of him are Louis, Hubert Sr., Gerard, Wiel, Johanna, and Joseph. Johanna, Gerard and Louis were the three surviving children from Hubert's first wife, Mina.)

The store: Look closely at the display (and fixtures), approximately 1906. (As the viewer looks into the store behind the glass door, one can barely see Hubert and Hubertina Dielen. In front of the store is Wiel Dielen on the bicycle.)

Three daughters and their parents in the meat store

Four sons and their father in the sausage factory

on with my persuasive oratory. Nevertheless, the general opinion was that I needed to make a better offer. So I gave my final offer of 13,000 guilders, which consisted of my entire savings, explaining all my cards had been put on the table and I could give nothing more.

As the Wolters family redeliberated my offer, Emile vigorously defended my position. Finally the Wolterses accepted my last bid, shook my hand, and the house became mine. The notary then inquired about the transferring of the deed. To which I said that was the family's business. After all, I did make it clear that that was all I had, nor could I scrape up anything else. The deliberation then began anew. For a moment there, I thought everything would yet come crashing down. Emile then said, "We have always been good neighbors and it is our desire for you to have the house rather than someone else, but the notary's expenses are for you, sir. So figure something out." I replied, the notary was their relative and could transfer the deed for a couple guilders. But very well, I'll gather the 100 guilders from my children's moneybox. The notary then wrote up the transfer documents stating the house was sold in its current state (as is), free of debt, for the price of 13,100 guilders and to be finalized with the deed registration and accompanying fees.

So that was an exhilarating day and my wife was ecstatic. The Singer (sewing machine sales) Company happened to be renting the house I had just bought and didn't mind relocating to my current residence. From then on, all of the preparations (to remodel the new site) were done in a timely fashion. I now traveled for a couple of days with Johanna to Köln, Bonn, Düsseldorf and Barmen-Elberfeld to study the latest in butcher shops. We then made a plan from which Henri Seelen (the architect) drew the final design. We bought the tiles from the *Hasselter Majolica Werke.* The windowpanes came from a glazier, in addition to some glass walls and shelves. The door was made by Frans van Dussen and the copper and lead encased windowpane was made by a Belgian named Huibrechts.

The majolica counter, the display window and other metals came from the Kamp brothers of Krefeld. By the time my cherished store was ready to open, it was quite a jewel. You could not find another store like it in the whole country. People even came from The Hague and Amsterdam to see my store. My family moved into the new residence as soon as possible and our former house was made ready for the Singer Company (renters). I now no longer needed my other small house (located next to Paulussen, across the street from me) on the Markt and after a thorough and prolonged negotiation with the mayor, the city bought it for 5,500 guilders. That was another fine sale.

In the meantime, I was frequently asked to accept candidacy as a city councilman but I always declined. That was until the butcher's guild had drawn up my candidacy of which I was then elected by a large majority. Once the war (WWI) broke out, I experienced a lot of hostility. Venlo was very anti-German and I was still considered a German even though I had been a naturalized Dutch citizen since 1897.

In Straelen, Louis had been sick for quite some time due to a stomach illness. As a result, by early 1914, Louis underwent surgery at St. Jozef's Hospital in Gladbach (Mönchengladbach). I visited Louis in the hospital a couple days after surgery and he appeared to be alright. On my second visit, he and I discussed how on Saturday he'd return to Straelen by automobile via Venlo so he could be with my family at noon. But on Friday, I was informed by phone Louis wasn't doing very well. So I caught the first train to Gladbach and when I arrived at St. Josef's Hospital, I found Louis wasn't in his room. A couple of Sisters (nuns) brought me to the mortuary and there in a provisional box under a cloth cover laid Louis, dead. After a while, I noticed someone standing next to me. It was Karl Theunissen, Louis's brother-in-law from Straelen. Louis had returned to Straelen Saturday by automobile all right, but not via Venlo. He had returned in a coffin.

IX. The World War (I)

Although my sons and I planned to not over slaughter in order to prevent excess inventory, we still ended up being overstocked with bacon, hams, and shoulder meats; which weren't in demand. In June, I went to the pig-market in Amsterdam where I bought some pigs for the price, after slaughtering, of 26¢/lb. As the mart came to a close, the customary auction took place and there I saw fine, firm, bacon sold for 18¢/lb. Not too long after Louis's death, the World War began. While the Germans still had a lot of money, they came shopping at my store and bought everything at any price. Thus, my store was able to sell its entire inventory at a good price.

My sons and I were now all expert butchers. I had four sons working with me and we accomplished quite a lot. We supplied the (Dutch) military for many years, which gave us a good income. My business was run mainly by my sons and daughters. Being a city council member kept me quite busy, especially since the new (city) slaughterhouse wasn't managed properly. Mr. Jules Thijnissen assisted me quite a lot and always gave me encouragement. Because we were both of German descent, we received a lot of opposition even though he and I were always of the view that the public interest took precedence, irrespective of any one person. But gradually more and more laborers and socialists became city council members and espoused theories I couldn't accept.

Rationing time had arrived and as president of the butcher's guild, I was often harshly criticized in the newspaper. Yep, and rudely insulted by Kobus Titulaer, even though mayor van Rijn and I had amply supplied the city of Venlo with meat, notwithstanding the flawed rationing.

I was also supplying the (Dutch) government with pigs that were readily available (from the farmers) here in Venlo. I had twice personally visited with minister Posthuma, who gave me much praise. But the real problem, however, lay in Venlo. Here, the butchers had been set up against me and no longer came to my assistance but rather thwarted me wherever possible. Once I realized all of my efforts weren't appreciated but met with lies, suspicions, and untruths; I resigned from the presidency and also ended my membership in the Venlo butcher's guild. That was to the great delight of butcher Schattorjé, who'd been unreserved in his encouragement to the opposition, and subsequently became the new president.

A restriction on slaughtering and some other things came about. Since I now had all of the butchers against me, things got a little difficult but I made sure my business got its fair share, member or not. Every household, butchers excluded, received a permit to slaughter one pig for their own private consumption. So off I went to the cattle inspector to ask for a slaughtering permit, which was denied. Then I appealed via telegram to minister Posthuma in The Hague. "Must feed 14 people. Are butchers excluded from slaughtering permits?" I received a reply in an hour. To: Dielen, Venlo, "Butchers are not excluded; slaughtering a pig for own use, permitted." Signed, Posthuma. With that, I returned to the inspector who chose not to receive me. But I made sure he was informed I now came with the approval of a higher authority. So I was then ushered in where I once again requested a slaughtering permit only to be refused in a surly manner. That was too much for me. I told him I had sanction from a higher authority. Now that he had to see! As he read the telegram, his countenance changed and with utmost courtesy, he now had the clerk write out a slaughtering permit. I, gentleman Dielen, was now graciously complimented as I was escorted to the door.

One of my sons then immediately went to a farm and bought a large, 500 lb. pig. The following morning, we slaughtered the pig to the great dismay of the bureaucrats and butchers. All sectors of the community wondered how I was able to get permission to slaughter the pig and I didn't volunteer anything either. After fourteen days, I again went for a permit with the explanation I still had 14 people to feed since my son previously only purchased a small piglet. The inspector now sardonically remarked that he was well aware of the cleverness of the Dielen's and didn't believe any of that. Nonetheless, he didn't hinder the writing out of another permit and reiterated that nothing of the pig was permitted to be sold. So we again bought a heavy sow, which we slaughtered and that provided us with plenty of meat.

But on Friday, there was a rumor all meat would be confiscated. I didn't believe that was ever going to happen, so I went for my usual glass of beer and game of cards at the Wilhelmina (an upscale café at the time). I played the card game skat with Thijnissen and Goossens, the baker. Also present was Herr Bieten, an attorney from Roermond. I asked him to take my place since I wasn't very good at the game and knew he was eager to join in whenever he was there. He mentioned that there was quite a stir in Roermond due to the rumor the butchers were going to have all of their meat confiscated. I couldn't understand how that course of action would even be considered. But the attorney said, "One couldn't think of anything so bizarre that The Hague is capable of contriving. My wife doesn't even have an ounce of meat in the house!" I told him my I encounter with the inspector and mentioned how I still had some meat provisions. After one more round of cards, we each went our own way as the attorney said, "Well mister, just make sure they don't seize your meat supply."

On Saturdays, my son Louis would return from Straelen. He'd help Aunt Stina in Straelen two days every week. Her sons were serving in the (German) military so Louis did the butchering and sausage making for her. Louis said the people in Straelen couldn't understand how the butchers (in the Netherlands) could continue to survive with the existing meat policy. Due to the existing meat policy, we decided to relocate our remaining meat supply to the garden-cabin at the Helbeek. We'd also station our meat-wagons at the garden plot. But a butcher who happened to be working on his own garden nearby must have noticed what we were doing because the following afternoon at 2 p.m., a couple of distribution officials and a regional constable, Houterman, came to inspect the garden-cabin. I accompanied and showed them the meat. They wanted to confiscate all the meat but I objected to that idea and asked, "On what grounds?" Then Mr. Welp, the director

The three oldest sons around 1915 (The word VLEESCHWAREN on the side of the delivery wagon means meat-products. The previous owner of the horse had been the crown prince of Germany.)

of distribution, also came by and said to the other officials they didn't have a search warrant. And so Mr. Wijnsbergen, himself an official, instructed his coworkers to leave. Just when I thought this whole crisis had blown over, an even higher-up official, Mr. Eusen, disagreed with Mr. Wijnsbergen and had the meat confiscated nonetheless. I asked him to leave the meat in the meat-wagon so as not to create a scene, but that he didn't do. He thought it would be good for the public to witness the exposure of a former city council member engaged in smuggling. I went home as the officials took the meat to the government warehouse, which was an old church on the Schriksel (a plaza).

Within half an hour, I was summoned to appear at the police station. There I encountered the chief of police, the regional constable, and some other officials. The police chief wanted to know whether the meat in the garden-cabin belonged to me. I answered in the affirmative and said I wanted it returned since no one had the right to confiscate my property. Now they all ganged up on me by spewing one falsehood after the other. I had had enough and said counter to the police chief, "I am Hubert Theodor Dielen and reside by the market plaza. If you want to speak to me, I'll be there. I'm no longer interested in arguing with you people about this." As I turned to leave, the police chief said, "No Dielen, you're under arrest and tomorrow you're going to Roermond (to be arraigned)." I was about to faint and asked for a doctor.

Dr. Janknecht came right away to examine me and immediately ordered I be discharged so I could have some bed rest. The police chief hesitated, but the doctor signed the medical certificate and that permitted me to go home. First, I had my son, Gerard, send a telegram to my attorney. Then Gerard held me by the arm as we walked home. We had to endure much derision as we waded through the swarm of people who had gathered in front of my house. A half-hour later, we were visited by an official who told me the attorney had ordered the chief of police to release me at once.

Early the following day, Gerard and Louis went to the attorney to explain everything, which included the telegram from Minister Posthuma. The attorney had them deliver 2 letters; one for official Eusen, saying the meat must be immediately returned, which he didn't do. And the other was for the commander of the provincial police in Venlo, with the injunction that if the meat hadn't been returned, he had to seize the meat from official Eusen and appoint Dielen as the caretaker. And so that is what actually occurred to the great dismay and consternation of the overruled officials.

A day later, my sons were seated with our attorney when the attorney had the regional constable Houterman and the chief (distribution official) Eusen summoned. There, those overzealous officials were censured and received a stern upbraiding. My sons returned home (by train) with greetings for father and the good news the meat would be returned before sundown. Gerard, after just arriving home, was notified to go to the government (distribution) office to retrieve our meat. They asked my sons to please keep it quiet from the public. To which Gerard reminded them how his father had the same request just a few days before and did they not remember their own conduct? Likewise, they shouldn't expect any discretion from us either. And so for everyone to see, the Dielen sons brought the meat home in great triumph. Thus were the gentleman government officials cured once and for all of harassing the Dielens.

Not much later, there came from The Hague a complete ban on the transporting of pigs with the exception of sows and pigs in heat. Here in Limburg, all mother pigs are pigs in heat. Whereas in (the provinces of) Holland, all of the pigs that were not used for breeding were castrated. The minister was obviously not aware of that. And so without saying anything, we promptly got into the pig trade for there was a lot of money to be had.

My son bought a pig for 270 guilders from Ton Willems in Schandelen (a former hamlet, today a suburb of Heerlen) and then sold the live pig for 645 guilders to W. van Oijen at the Markt (the plaza). The pig weighed 215 kilograms and was sold for 3 guilders per kilogram. Thus, we cleared a profit of 375 guilders from one pig. That was a first! Since all of the pigs had to be turned over to the government, no one could get any with the exception of Dielen; and that at a reasonable price.

The government would give 2½ guilders per kilogram for a live pig. Dielen's pigs were only for breeding of course; otherwise Dielen wasn't permitted to trade in pigs (for slaughtering). But if a pig had an injury or discontinued eating (a sign of illness), then the mayor would always grant an emergency slaughtering-permit. The mayor knew what was going on and that was fine

Hubert Dielen with his family on his 60th birthday in 1915. (Front row from the left: Regien, Hubert, Petronella, Jean, Hubert Sr., Hubertina, and Marie. Back row from the left: Wiel, Johanna, Joseph, Gerard, and Louis.)

with him, as long as I made sure that Venlo had a sufficient supply of meat.

But there was of course still the perpetual jealously to deal with. And on one particular day, all of a sudden a gentleman stood in front of me. He asked whether I had any pigs. I replied I did indeed have pigs, three, in fact. They were in the cellar and were only for breeding of course. I'd be willing to sell him one, but slaughtering was out of the question. At that moment, he then presented himself as inspector Schaap of The Hague; to which I said, "I hope you know something about pigs," as I escorted him to the cellar. I remarked how the pigs even looked quite gravid but didn't say they also had just gorged themselves. I then noticed to my horror one of the three pigs was actually a boar. Then the fellow asked, why I had permitted E. van der Loo of Villapark to slaughter a pig? To which I replied how it would've been a pity to simply let the poor animal waste away just because it was sick with a swelling in its neck.

As the inspector left, he said pertaining to the Van der Loo pig, I'd be hearing more. The animal had been seized and temporarily impounded at the (city) slaughterhouse. That evening, I heard the inspector telephoned The Hague and was told he'd receive assistance the same evening. At 9 p.m., I went to the Wilhelmina for an hour of card playing. There, the innkeeper told me something was in store for me the following day as he pointed out the 4 gentlemen from The Hague.

Sure enough, the following morning, a Saturday, 4 fellows were at my door wanting to speak to me. I introduced myself and asked them to come in. I asked them if they'd like coffee, tea or cognac with a cigar, but they declined my offer. The oldest fellow, a man of about 60 years of age, said he was the director for the provisioning of pigs and wanted to know whether I had delivered pigs to private individuals in Venlo and whether I still had three pigs in the cellar, to which I answered yes to both questions. I offered to show them the cellar. I told the gentlemen if they weren't well experienced with pigs, we could just as well stay upstairs. To which the director asked why I had to say that, since he claimed they were also butchers. I mentioned how the foolish inspector sent by The Hague was undoubtedly an Israelite and everyone in Venlo was daily reminded in the newspaper how Israelites/Jews didn't know anything about pigs (because Jews had to adhere to strict kosher laws). I said if you fellows were also Israelites (Jewish), there would be no point in going to the cellar. Deeply offended, they went down to the cellar where I let them see everything. There, I had converted the salting pen to a pigpen, since we (butchers) were no longer allowed to slaughter pigs. It didn't seem to dawn on these fellows either that they were looking at two sows and one boar. They did compliment me though on having a fine (salting) cellar. Back upstairs, the director asked me who gave me the permission to keep pigs. But I wanted to know since when was it illegal and by whom? Then he said they had come to confiscate every pig. First they seized Van der Loo's animal, which was destined for the sausage factory in Boxmeer and now they've come for my pigs. I told them they better not seize my animals and send them to Boxmeer for they would most surely have to return them. "Just a moment, I'll give the mayor a call." I intentionally left the door open so they could easily hear everything. "Mayor? This is Dielen. I have 4 gentlemen here from The Hague who are telling me they are confiscating all pigs and sending them to Boxmeer! Will you make sure none of the pigs are removed until I've had a word with you? It seems the gentlemen here don't know their own laws, just as you've always said." The mayor instructed me to have the gentlemen stop by his office when they were ready to leave.

Then the director asked me on what basis did I have for such contentiousness? But I wanted to know on what grounds did he have for his version of things? "From the last circular of the Minister," he said, as he pulled out his paper. I then pulled out the exact same circular from the papers in my bureau as the director began reading out loud. As we came to the relevant passage, "It is illegal to transport all pigs, with the exception of boars, sows and pigs in heat." "Stop," I called out. "That's it. Gentlemen, I'll now clear up your misunderstandings so you'll be home by tonight. The Minister doesn't know that here in Limburg, the mother pigs are not neutered, thus they are all pigs in heat and are allowed to be transported. Therefore, these are all breeding pigs, and if an illness or injury is discerned, as with the Van der Loo pig, then we perform a mercy slaughter. Thus, nothing illegal occurs, but rather you simply have an inadequate law." I then asked the gentlemen whether they were now ready for a cup of coffee or a cigar, to which they now readily accepted. They asked me whether I wanted to pick up the Van der Loo pig. That I would do with pleasure providing they compensated me 5 guilders. Also, they shouldn't forget to pay the mayor a visit. When I then asked them what should I now do with my pigs? They said, "Breed 'em!"

As the men went on their way toward the city hall, I quickly telephoned the mayor to inform him of the latest developments. He said, "I'll be waiting for them and receive 'em in such a manner they won't forget me for quite some time." While the gentlemen were with the mayor, I strolled around the plaza until I saw the 4 gentlemen leave the city hall, red with embarrassment. As soon as I had oriented the gentlemen toward the train station, I politely bid them farewell. To my astonishment, I noticed a couple of butchers standing on one of the street corners obviously looking to see this latest outcome of events. I later heard those butchers had already begun a rumor that my pigs had been confiscated. They were, of course, very disappointed when they realized Dielen could still keep his pigs. Then when the mayor came down the steps, shook Dielen's hand and loudly laughed, the butchers dispersed.

Soon the pig business came to an end and we switched over to chicken, ducks and geese. But that messy job was quickly abandoned. Although we still had quite a lot of provisions in the house, my sons would go to the farmers to purchase bags full of wheat. We had bought a small wheat mill and now could grind out enough flour to make our own bread. Thus, we got through the war period without serious scarcity.

X. The Children

On October 3, 1916, Hubert (who went blind around 1910) married Josephine Selen. They rented butcher Kölm's house at Heutz Street, Venlo with the option to buy and 4 years later, they did just that. Willy, who was engaged to a Marie Peeters (from Venlo, she had been adopted), took over J. Hillen's business in Zaandam, also with the option to buy later. Then Louis, who was engaged to a Marie Peeters from Roermond, decided to get married. But since there were no rentals available, he ended up buying an overpriced house at Neer Street, Roermond from a Frans Coolen. (Louis later began a butchery in Heerlen but switched over to a concessionaire there that was owned by the national mining company). Then came Regien, who married Hubert Selen (the brother of Josephine Selen). (Hubert and Regien ran a large greenhouse vegetable business). After them, Gerard ended up getting married to a Wilhelmina Beurskens. Then came Nellie, who married Jacques (Jac) van Dael. I bought for Nellie a brand-new house at a good price in Swalmen. (They ran a pastry/confectionary shop with an adjoining lunch/tearoom, which was later expanded into a hotel and restaurant.)

Marie, who had been for 6 years at a boarding school in Grubbenvorst (which was run by the Ursuline Sisters), was sent to a clothing design/tailor school in Amsterdam. Once she got her diploma in tailoring, she briefly came home and then went on to the Ursulines convent/school (near the Venlo border with Germany). After a short while, she decided to become a nun but ended up becoming seriously ill. I then had to bring her home.

The doctor, who had given Marie a thorough examination, was mystified. Due to the excellent nursing care from her mother, Marie got well. Against my wishes, Marie returned to the Ursuline school, but this time to teach. That went all right for a couple of years until she again became ill and weak. She again came home and slowly recuperated. When she entertained the thought of going back to the school, I told her, "Absolutely not."

Jean attended a school in Weert called "Bishop's College." After that, he went to a similar school in Rolduc. Then he went to a business school (in Wageningen) where he received a 5-year diploma and then worked in an office for an ironmonger in Terborg for 3 years. Then he became an editor of *The Maasbode,* a well-known Catholic newspaper in Rotterdam. (He married Dorien Linskens from Rotterdam. He later became the chief advertising manager of a weekly agricultural magazine. His oldest son, Jan, is the compiler of this book in Dutch.)

Joseph began working first as an apprentice for 3 years at H. Bloem's pastry/confectionary business. Then he went to the provinces of North and South Holland, Belgium, and France. He worked as a cook and pastry/confectioner at the best hotels in the major cities there before serving on the Holland America Line where he made 3 tours to America. Then he signed on with the liner Netherlands Lloyd and made 2 tours to Indonesia. Returning from all of that, he wanted to start his own business in Venlo but wasn't able to find just the right house/building. So he went instead to Amsterdam where he again worked as a pastry/confectioner. Then he worked again as a cook, but this time on *The Wagon-Lits* (a first class wagon-lit). He was now 29 years old and wanted at all costs to begin his own business. Then a little while later, I heard that the Joosten hotel, owned by Wiel and Pien Wolters, was up for sale. So we purchased the building and renovated it in an upscale manner into a hotel, restaurant, and tearoom/pastry store. Since this was a joint enterprise by both Joseph and Johanna, no expense was spared. The new business was inaugurated as the Hotel American Lunchroom. Then Johanna had a change of heart and withdrew as joint owner, but nonetheless did continue to give financial and management advice. Gerard got married in 1922 and since the other three sons already owned a butcher business, it was agreed upon that Gerard and Johanna would inherit my business (not the house). After all, Gerard and Johanna had always been working at home.

The main reason I lost interest in running my business was that in these last years, just like before the war, we weren't saving any money. In fact, we were losing money. I felt that the company, of which I originated and was responsible for (sole proprietorship), no longer consulted me and spent too much money. Although I recommended a hundred times, no, a thousand times, that they heed my suggestions, that didn't occur. Their recalcitrance was such that when I'd voice my objection, they just went ahead and did things their own way. That became unbearable for me. Consequently, financial capital was lost due to poor preparation of the meats. It perplexed me that I, the experienced butcher with great insight, was unable to teach them anything as long as they were working for me. They never took my advice or exactly followed my method, never. Consequently, a lot of money was wasted. That's why I converted the business into a partnership in 1920. At first, it didn't make any difference. But now it seems that Johanna and Gerard are making a decent profit, which greatly pleases me. After all, they have to pay me 60 guilders per week in rent.

When I purchased the Wolters residence, I was originally going to combine that building with my own to make one large business. Then when my sons got married, I would open branch businesses at Roermondse Poort (Roermond Gate), Keulse Poort (Köln Gate) and Gelderse Poort (Geldern Gate) (all are entrances/exits of Venlo). The business here on the Market Plaza would be the hub that supplied the branch businesses. I'd then appoint each of my sons to be manager/owner of their respective branch in which the wife would actually run the branch store while the husband worked (during the day) at the hub. My oldest son would reside at the hub and oversee everything. When the first son to get married simply opened his own business, it dawned on me that no one really took my idea all that seriously. Hence, my grand plan came to naught. I thought that quite a pity, since I had anticipated many a time of being the C.E.O. of the company H. Th. Dielen. Then the city of Venlo would be able to see what the Dielen family could really do. But oh well, we can now say, "It would've been nice, but it wasn't meant to be."

But I must also thank The Good Lord everyday that he has helped us so often and none of our well-behaved children ended up debauchers or drunkards. So if anyone now asks, "Why didn't Hubert's children follow his advice?" I say it was my own fault of which I'll do my best to explain. When I married my 2nd wife, she always had the fear that the first 3 children would only consider her a stepmother; and due to her good-natured disposition, she gave the (oldest) children everything they wanted and tolerated a lot just to keep them satisfied. So long as the children were young, it didn't matter much and I also enjoyed her obliging manner. All of my children got along so well together that when the first 3 were around 20 years old and the subject of their having a different mother from the younger children was brought up, they didn't think it a big deal. Thus, years later were they able, without any problems, to make an agreement they'd all think of themselves as equals amongst one another.

Everything went great the first years (after the agreement) when everyone was still cooperating. But then some discord began to appear. At first there were just some trivial disagreements, but then larger differences of opinion arose. For the sake of preserving the peace, I didn't say much at first, but things went from bad to worse. When I saw things that were not up to standards, I had to raise my voice. But then Momma, kind hearted as she was, would always intercede on the side of the children. Although at times I complained or appeared heavy-handed to Hubertina, the disharmony compounded and continued without letup. I was always considered in the wrong. It wasn't long before whatever I said was already disapproved of before I had even finished my sentence. So I then decided if kind words weren't appreciated, I'd simply demand they follow my instructions; which is what I tried at one time but was quickly cured of that idea. My wife and children would always side against me. I'd end up the good-for-nothing and she would be the good woman. Then there would be a little crying and I'd give in. That was where I made my big mistake. I should've stood firm, come what may; and because I didn't stand my ground, I, myself, am mostly to blame for all my grand plans going amiss. I should've been more insistent and put my foot down. What is done cannot be undone and we must thank The Good Lord everyday that everything has nonetheless turned out just fine.

To date, I have given to each of my children that have either gotten married or started a business, 6,000 guilders for a trousseau and/or working capital without having to compensate me.

Epilogue

The real work was done by my daughter Martina, who did the typing and her husband, Ronald Lips, who did the layout.

Jan Dielen

The Transcriber's Epilogue

I want to thank Ken Fowler, my indispensable computer expert, for his computer assistance and creation of the necessary maps. The book wouldn't be complete without those maps!

I also want to thank Donna Pacheco of Olympic Printers, Inc. in Port Angeles, WA for all of the computer assistance she has provided; and I want to acknowledge Cindy Clardy for her computer assistance, especially the adding of 7 villages to the map on page x.

Some more thank-yous go to Kristen Butler of Composition Services at BookMasters, Inc. for reformatting the 4th edition of the book, to Ryan Feasel for redesigning the cover, and to Emily McQuate—my account executive at BookMasters, Inc.

Mike Reininger

Glossary

KOLPING. In order to present one's self as a skilled craftsman in Germany, one had to be an apprentice for 3 years and then another 7 years as a journeyman which often involved transferring from one boss to another as one journeyed. That was called *wanderzug* (going on a long trek). Adolf Kolping, who as a journeyman shoemaker had personally experienced the miserable circumstances under which journeymen had to live, decided to do something about it. Adolf studied for the priesthood and once ordained, he dedicated himself entirely to the establishment of *Kolping homes,* an organization still in existence, which was especially created for traveling journeymen. One could usually spot a traveling journeyman by his *Berliner,* which contained his most essential clothing items, etc.

KIEP. A straw basket one carried on the back. It is still in use for grape harvesting in some wine regions. An example of the basket can be seen on page 132.

UHLANEN. Cavalry. An Uhlanen regiment, a sort of occupation force, was stationed at Thionville, Alsace-Lorraine, from 1870–1918. Those two provinces had been annexed by the Germans after the 1870 Franco-Prussian war and then returned to France after WWI.

Genealogy
(added by editor)

Hubert Theodor Dielen marries Wilhelmina Engels 1st and then Anna Maria Hubertina Engels.
Born: 1/28/1855 Dies: 4/6/1926 Born: 11/12/1852 Dies: 1/10/1890
Born:11/1/1857 Dies: 8/7/1941

The August 1937 family photograph was done in honor of Anna Maria Hubertina Dielen's 80th Birthday.

1st Johanna, born: 7/6/1884, died: 5/15/1971–Chose to be absent from the August 1937 photograph.
2nd Gerard, born: 4/18/1886, died: 2/6/1960. Married: Wilhelmina Beurskens.
- 1st Wilhelmina (Mien) Dielen b. 1923
- 2nd Tina Dielen b. 12/26/1924, d. 5/11/1991 (married a successful land developer divorcée. They had two daughters.)
- 3rd Trūs Dielen b. 1926
- 4rd Hubert Dielen b. 8/15/1927, d. 11/9/1993
- 5th Louis Dielen b. 1929
- 6th Gerard Dielen b. 1930
- 7th Anneke Dielen b. 1932
- 8th Jack Dielen b. 1936, (had been managing director of International Group Companies. He also was a member of an advisory board for the Royal Dutch Industries Fair (Jaarbeurs), chairman of Association of Technology Process Control of Food & Chemicals, and chairman of Knowledge Center Universities & Trade and Industries.

Due to his accomplishments, he was knighted by the queen. He has a son and daughter.)

3rd Louis, born: 5/22/1887, died: 11/12/1970. Married: Maria Peeters.

1st Hubert (Ber) Dielen b. 1922

2nd Wilhelmina (Mien) Dielen b. 1923

3rd Johanna (Jo) Dielen b. 1925

4th Gerard b. 8/1/1927, d. ?, absent from August 1937 family photograph.

5th Mia Dielen b. 4/5/1930, d. ?

[4th Hubert, born: 9/29/1888, died: 8/27/1890]

[5th Wilhelmina, born: 1/10/1890, died: 8/19/1890]

4th Hubert Gerard Christian (went blind around age 19), born: 2/17/1891, died: 5/13/1984. Married: Josephine Selen.

1st Hubertina (Tina) Dielen b. 8/1/1917, d. 10/31/2006, married Piet van Dongen. They had 9 children.

2nd An Dielen b. 9/13/1918, d. 5/5/2006

3rd Catherina (Tōs) Dielen b. 10/5/1919, d. ?

4th Marie (Mieke) Dielen b. 12/12/1920, d. ?, (she lived in South Africa when she was married)

5th Tony Dielen b. 7/18/1922, d. 3/1/2002

6th Regien Dielen b. 1924

7th Hubert (Hūb) Dielen b. 7/22/1926, d. ?, (managed the snack bar at a Venlo tennis club.)

8th Jan Dielen b. 1928

9th Joep Dielen b. 4/11/1930, d. 5/24/2004, (owned a wine store in Venlo, has two sons and two daughters.)

10th Gerard Dielen b. 1931

11th Wiel Dielen b. 10/15/1932, d. ?, (died in South Africa from a diving board accident late 1950's)

12th Anton Dielen b. 1934, (managed a tennis court business in Venlo.)

13th Frank Dielen b. 1965 from 2nd wife

5th Wilhelm Lüdwig Maria (a.k.a. Willy), born: 2/28/1892, died: 2/17/1970. Married: Marie Peeters (adopted).

1st Wilhelm (Willy, Jr.) Dielen b. 4/6/1922, d. 12/28/1987

2nd Hubertina (Tina) Dielen b. 1923

3rd Antōn Dielen b. 3/8/1925, d. 9/2/1983

4th Catherina (Tōs) Dielen b. 1928

5th Clara Dielen b. 1934

6th Joseph, born: 11/6/1894, died: 6/9/1952 in Velden (across the river from Grubbenvorst). Married: Maria Dekkers.

1st Maria (Miep) Dielen b. 4/14/1930, d. 2/20/1994, married Jack Rose (very successful architectural engineer) of Memphis, Tennessee. They had two daughters, Michele and Monique, and finally lived at Vero Beach, Florida.
2nd Theresa (Resie) Dielen b. 1933, (Resie and her husband are avid bird-watchers.)
3rd Josephine (José) Dielen b. 11/15/1934, d. 10/5/1994
4th Joseph (Jōs) Dielen b. 1938, (engineer).
5th Wiel Dielen b. 1942, (owns a successful catering business in Birmingham, Alabama.) Married Ann Templeton and has a daughter named Ann Marie and a son named Wiel.
6th Hans Dielen b. 1946, (car dealer, 1st of Volkswagons and then of Porsches)
7th Yvonne Dielen b. 1951, (Fine artist, currently resides in Vienna.)

7th Petronella (Nel) Josephina Regina Dielen, born: 3/3/1896, died: 6/5/1983 in Venlo. Married 10/1/1923 to: Jacobus (Jac) Hubertus van Dael, who was born: 4/13/1891 in Swalmen, Netherlands. Died: 8/12/1931 in Roermond of tuberculosis. He was a baker/pastry chef. The van Daels were sextons in their church from 1778 to 1956. Jacobus's eldest brother, Johannes Hubertus van Dael served from 1896–1939 and was awarded 12/6/1936 for his 40 years of service the golden cross of honor, "Pro Ecclesia et Pontifice," from Pope Pius XI.
1st Hubertus (Ber) van Dael b. 6/28/1924, d. 11/25/1983 (was an airline steward/purser for KLM)
2nd Hubertina (Tina) Petronella Johanna van Dael b. 1925 in Swalmen, (Translator of Hubert's autobiography.) Married Thomas Reininger (government service worker) of Minneapolis, Minnesota and had a daughter and five sons. Michael, her 5th child, edited the English version of Hubert Dielen's autobiography. Tina and Mike live in Sequim, WA. (Robert E. Reininger, her oldest son and 2nd child, is a retired captain of the U.S. Coast Guard.)
3rd Martha b. 2/8/1928, d. 6/29/1989

8th Regina (Regien), born: 8/31/1897, died 1/22/1985 in Venlo. Married: Hubert Selen (brother of Josephine Selen mentioned above).
1st Jan Selen b. 1924, (successful personnel manager)
2nd Hubert Selen b. 1928, (knighted in the 1990's for his efforts in bringing world gliding championship events to

the Netherlands. His son even won a world gliding championship.)

3rd Gerard Selen b. 1930

9th Maria Hubertina Beatrix, born: 2/4/1899, died: 9/8/1978—chose to be absent from the August 1937 photograph.

10th Johannes (Jean) Hubertus Philippus, born: 5/26/1900 in Venlo, died: 6/18/1957 in Voorburg (near The Hague). Married: Dorien Linskens.

1st Jan Dielen b. 1931, (successful purchasing manager and compiler of Hubert Dielen's autobiography.) Married Riet Steegers and has two daughters, Hanneke and Martina.

2nd Stephen Dielen b. 11/2/1934, d. 1/3/1993

3rd Catherina (Tōs) Dielen b. 1936, married Bernard Geradts (successful fertilizer salesman) and has three daughters: Dorine, Henny, and Bernice. Tōs and Bernard live in Tampa, Florida.

4th Karel Dielen b. 1937

5th Gerard b. 1943

6th Anton (Tony) b. 1945

o Danielle,
njoy the book!
rom,
mike

Happy reading,
Danielle
Best wishes,
Tina

Spring 1984–Michael and Hubertina Reininger with Regien, the last of the old-timers.

Jan Dielen 11/02/2003

Dielen Family Photograph-August 1937

August 1937 Family Photograph in Venlo

Top row left to right
1. Hubert, 4th child of Gerard
2. Jan Dielen, 8th child of Hubert
3. Regien Dielen, 6th child of Hubert
4. Tina van Dael, 2nd child of Nel and the translator of Hubert's autobiography
5. Tōs Dielen, 4th child of Wiel.
6. Tony Dielen, 5th child of Hubert
7. Hubert Dielen, 7th child of Hubert
8. Johanna Dielen, 3rd child of Louis
9. Hubert Selen, 2nd child of Regien
10. Trūs Dielen, 3rd child of Gerard
11. Antōn Dielen, 3rd child of Wiel
12. Hubert van Dael, 1st child of Nel

2nd row left to right
1. Tina Dielen, 1st child of Hubert
2. Mien Dielen, 2nd child of Louis
3. Mieke Dielen, 4th child of Hubert
4. Hubert Dielen, 1st child of Louis
5. Tina Dielen, 2nd child of Wiel
6. Willy Dielen, 1st child of Wiel
7. An Dielen, 2nd child of Hubert
8. Tōs Dielen, 3rd child of Hubert
9. Jan Selen, 1st child of Regien
10. Mien Dielen, 1st child of Gerard

3rd row left to right
1. Clara Dielen, 5th child of Wiel
2. Marie Dielen-Peeters, wife of Wiel
3. Hubert Selen, husband of Regien Dielen
4. Maria Dielen-Peeters, wife of Louis
5. Wiel Dielen, 2nd child of Hubertina Dielen-Engels
6. Gerard Dielen, 2nd child of Wilhelmina Dielen-Engels
7. Louis, 5th child of Gerard
8. Louis Dielen, 3rd child of Wilhelmina Dielen-Engels

4th row left to right

1. Jack Dielen, 8th child of Gerard
2. Mien Dielen-Beurskens, wife of Gerard
3. Hubert Dielen, 1st child of Hubertina Dielen-Engels
4. José Dielen, 3rd child of Joseph
5. Marie Dielen-Dekkers, wife of Joseph Dielen
6. Jean Dielen, 7th child of Hubertina Dielen-Engels
7. Nel van Dael, widow, 4th child of Hubertina Dielen-Engels
8. Hubertina Dielen-Engels, 2nd wife of Hubert Dielen, Senior.
9. Stephen Dielen, 2nd child of Jean Dielen
10. Regien Dielen-Selen, 5th child of Hubertina Dielen-Engels
11. Joseph Dielen, 3rd child of Hubertina Dielen-Engels
12. Josephine (a.k.a. Fien) Dielen-Selen, wife of Hubert Dielen and sister of Hubert Selen
13. Dorien Dielen-Linskens, wife of Jean Dielen
14. Tōs Dielen, 3rd child of Jean Dielen

5th row left to right

1. Joep Dielen, 9th child of Hubert
2. Mia Dielen, 5th child of Louis
3. Jan Dielen, 1st child of Jean Dielen and the compiler of Hubert's autobiography
4. Tina Dielen, 2nd child of Gerard
5. Martha van Dael, 3rd child of Nel
6. Wiel Dielen, 11th child of Hubert
7. Resie Dielen, 2nd child of Joseph
8. Gerard, 6th child of Gerard
9. Anneke Dielen, 7th child of Gerard
10. Miep Dielen, 1st child of Joseph
11. Gerard Dielen, 10th child of Hubert
12. Anton Dielen, 12th child of Hubert
13. Gerard Selen, 3rd child of Regien

The Dielen family tree of middle and North Limburg

To date, our Dielen ancestry has been traced back to a farmer on the property tax roll of 6/24/1582 in the small village of Merselo (near Venray). He had a grandson named Godefridus.

Godefridus Dielen was born circa 1620 and died 1/20/1669 at the village called Well.
The neighbors called the children by the father's nickname Geurts. The oldest child's name was Wilhelmus Dielen. Then there was Johannes, Henricus, Petrus, Gertrudis, and Maria.

Wilhelmus Dielen married Vita Cocken from Höckelom (Germany).
Born: circa 1660 Born: circa 1665
Oldest child was born 1690 and named Godefridus. Then came Henricus who was born 6/5/1696. The 7th of the 10 children, who were all born in Well, was Matthias (Theodorus) Dielen.

Matthias was born 12/27/1706. Married 9/11/1730 at Blitterswijck, County of Geldern, Prussia to Anna Coopmans Mulders of Blitterswijck. She was the daughter of Petrus Coopmans Mulders and Wilhelma Muysers of Blitterswijck. All of Matthias and Anna's 11 children were born in Blitterswijck.

The youngest child was Martinus Dielen. Born: 4/18/1752. Married 10/29/1781 to Aldegundis from Well, County of Geldern, Prussia at Broekhuizen, County of Geldern, Prussia. She was born in Arcen, County of Geldern, Prussia 10/5/1754 and died

5/26/1829 at Broekhuizen, Limburg, Netherlands. He died 10/25/1815 at Broekhuizen, Limburg, Netherlands. She was the daughter of Joannes of Well, who in turn was the child of Venraij and Maria Roosen. Martinus and Aldegundis had 7 children.

The oldest was named Joannes Mathias Dielen. Born: 10/17/1782 in Broekhuizen, County of Geldern, Prussia. Married for the 2nd time in 1/29/1815 at Broekhuizen, Limburg to Johanna Wijnen, who was born 3/14/1789 in Kessel, Duchy of Cleves, Prussia/Germany. She died in Venlo, (East) Limburg, Belgium 1832. She was the daughter of Wijnandus Wijnen and Maria Meertens of Kessel. Joannes Mathias Dielen was a tanner and glove/mitt maker. He died in Venlo, Limburg, Netherlands 1/4/1826. They had 3 children: Johannes Martinus Dielen who was born 11/11/1815 in Broekhuizen, Limburg, Netherlands; Wijnandus, who was born 1/30/1819; and Marie (a.k.a. Mieke Tante), who born circa 1922 and died 1/19/1900.

Johannes Martinus Dielen was a butcher. He married in 10/23/1845 to Regina Hubertina Brockmann, who was born 11/11/1821 in Straelen, Westphalia Prussia. He died 11/23/1884 in Straelen, Germany. She died 7/25/1880 in Straelen. They had 5 children: Johanna Hendrika, who was born 4/28/1848 and died in 1901; Lüdwig (Louis) Mathias Dielen, who was born 9/18/1850, worked as a butcher, and died 3/25/1914; Peter Johan (Johan Jr.), who was born 7/23/1852, worked for the R.R., and died in 1879; *Hubert Theodor Dielen,* who was born 1/28/1855 in Straelen, Westphalia Prussia and died 4/26/1926 in Venlo, Limburg, Netherlands; Philip Ignaz, who was born 7/31/1857, worked as a baker, and died 7/13/1885.

The parents of Hubert's wives were: Gerard Engels—candymaker, butcher, café manager/(owner?) from Venlo who died in February 1897 of a heart attack; and Beatrix Hendrickx who died circa 1867. The Engelses had around 8 children. Wilhelmina was the oldest who was born 11/12/1852, married Hubert 10/2/1883, and died 1/10/1890; then perhaps Jean who would've been born circa 1854; Martinus, who was born in 1855 and died of alcoholism 12/11/1895; Anna Maria Hubertina, who was born 11/1/1857 in Venlo, married Hubert 7/28/1890 and died 8/7/1941; Nel; Maria; one other daughter married to Mattousch; and then Gon, the youngest daughter.